P9-AGQ-057

About Island Press

Island Press is the only nonprofit organization in the United States whose principal purpose is the publication of books on environmental issues and natural resource management. We provide solutions-oriented information to professionals, public officials, business and community leaders, and concerned citizens who are shaping responses to environmental problems.

In 1999, Island Press celebrates its fifteenth anniversary as the leading provider of timely and practical books that take a multidisciplinary approach to critical environmental concerns. Our growing list of titles reflects our commitment to bringing the best of an expanding body of literature to the environmental community throughout North America and the world.

Support for Island Press is provided by The Jenifer Altman Foundation, The Bullitt Foundation, The Mary Flagler Cary Charitable Trust, The Nathan Cummings Foundation, The Geraldine R. Dodge Foundation, The Charles Engelhard Foundation, The Ford Foundation, The Vira I. Heinz Endowment, The W. Alton Jones Foundation, The John D. and Catherine T. MacArthur Foundation, The Andrew W. Mellon Foundation, The Charles Stewart Mott Foundation, The Curtis and Edith Munson Foundation, The National Fish and Wildlife Foundation, The National Science Foundation, The New-Land Foundation, The David and Lucile Packard Foundation, The Pew Charitable Trusts, The Surdna Foundation, The Winslow Foundation, and individual donors.

About SeaWeb

SeaWeb was launched in 1996 to raise awareness about the growing threat to the ocean and its living creatures. SeaWeb's goal is to inspire a new and vigorous commitment to protect the global marine environment that results in a public call for stronger and more precautionary policies aimed at protecting oceans worldwide.

Believing that the best tool to protect the ocean is knowledge, SeaWeb strives to make credible scientific information about the ocean environment accessible to the public. The organization's main function is to educate, which it does through a variety of communication outlets. Public awareness of and concern for the ocean is monitored through public opinion polling. SeaWeb sponsors and produces educational programming on radio, television, film, and the Internet, and organizes delegations of scientists to meet with journalists, editors, and other news media professionals. SeaWeb also produces a great variety of written materials to circulate to the public and news media. With the help of scientists, educators, researchers, and communications specialists, this organization has become a respected, independent resource for journalists, government officials, students, and concerned citizens.

The Living Ocean

THE LIVING OCEAN

Understanding and Protecting Marine Biodiversity

SECOND EDITION

Boyce Thorne-Miller

Foreword by

Sylvia Earle

SeaWeb

ISLAND PRESS

Washington, D.C. • Covelo, California

The author gives grateful acknowledgment to Alice Jane Lippson for the use of her illustrations, which appeared originally in *Life in the Chesapeake*, 1984 and 1997 (2nd ed.). Illustrations copyright © 1984, 1997 by Alice Jane Lippson.

Library of Congress Cataloging-in-Publication Data
Thorne-Miller, Boyce.
The living ocean : understanding and protecting marine biodiversity / Boyce Thorne-Miller ; foreword by Sylvia Earle. — 2nd ed.
p. cm.
Includes bibliographical references (p.) and index.
ISBN 1–55963–677–7 (cloth). — ISBN 1–55963–678–5 (pbk.)
1. Marine biology. 2. Biological diversity. I. Title.
QH91.T45 1999 98–42545
333.95'616—dc21 CIP

Printed on recycled, acid-free paper

Manufactured in the United States of America
10 9 8 7 6 5 4 3 2 1

Contents

Foreword

> We know now what was unknown to all the preceding caravan of generations, that men are only fellow voyagers with other creatures. . . . This new knowledge should have given us, by this time, a sense of kinship . . . a wish to live and let live; a sense of wonder over the magnitude and duration of the biotic enterprise.
>
> —Aldo Leopold, *A Sand County Almanac*

Given the task of designing creatures suitable for living on Earth, first priority surely would be placed on features such as fins and gills, on the ability to communicate by means of light and sound—on characteristics, generally, that would enable an organism to survive in an aquatic medium. After all, as any astronaut can tell you, Earth is an ocean planet, an exceptional place in the vast reaches of space where water—the single nonnegotiable requirement for life—is wondrously abundant. Most terrestrial primates seem blissfully unaware that 97 percent of Earth's water is salty, cold, and dark—and that the sea makes up about 97 percent of the biosphere, the lively, habitable part of the planet that is the focus of this most timely, vitally important book by Boyce Thorne-Miller.

Often, when I dive into the ocean, I imagine myself plunging into the history of life on Earth—a metaphor made real as I am surrounded by small beings whose ancestry is linked to the great explosion of biodiversity that occurred hundreds of millions of years ago. Dinosaurs are symbols of ancient times, of a fascinating era that long preceded the appearance of humankind. Yet as I glide over a coral reef, wend my way among slippery strands of kelp, or descend into a deep submarine canyon, all around me are creatures that would be familiar if the ocean were today as it was 400 million years ago—long before the first dinosaur appeared.

Most of the major divisions of animals, plants, and microbes that have shaped the character of Earth from the beginning of life to the present time are still around, bobbing or gliding or pulsing or swimming in the stuff of life itself—water. Representatives of about half of the broad categories of living things have ventured onto land, but all life on Earth—whether in arid deserts or in lush rain forests—is inextricably bound to the sea. Water is the key. Most of the rain, sleet, hail, and snow that falls anywhere on Earth originated as mist rising from the sea into clouds—a link as vital to the way the world works as is the continuous flow of water from the land back to the sea via streams, rivers, and underground trickles and seeps.

But it isn't just rocks and water that form the basis for making Earth a suitable place for humankind and the rest of life as it now exists. As this book makes clear, living creatures, acting on the basic ingredients present when the planet was formed, have shaped the nature of the place for billions of years, and they shape it still. Oxygen generated by microbes ages before there were multicellular organisms paved the way for the next phase of diversification. Then, as now, bodies of the quadrillions of creatures living in every square mile of energy-fixing sunlit surface waters settled deep in the sea, storing carbon and socking away nutrients that later resurfaced in upwelling currents, triggering cycles of growth that in turn caused new patterns to form. The system is ever changing, but for thousands of millennia the shifting patterns have existed within a framework that allows life to prosper—even when asteroids thundered onto the scene; even when pulses of volcanic action covered vast areas with clouds of dust and ash; even as ice ages came, gave way to warmth, and came again.

An underlying and unanswered question provoked by Thorne-Miller's thoughtful treatise concerns the outcome of humankind's effects on life in the sea—effects grave enough that some equate them with whatever natural disaster caused dinosaurs and much of the rest of life on Earth to die 65 million years ago. Time will tell, of course, but rather than wait and hope that no matter what we do, the resilience of nature will somehow enable the planet to carry on within limits that suit us, maybe we should be inspired to take action now to protect and maintain what we can of the living systems that sustain us. This, perhaps, is the most significant message of *The Living Ocean*.

Ten thousand years ago, our early ancestors did not have to worry about such weighty matters, blessed as they were with a planet emerging from the most recent age of ice. For a while, it was possible for us to prosper by hunting and gathering, much as other wild creatures do, carving out a place for ourselves among the thousands of other species that shared our range. On land, as our numbers grew, it did not take us long to diminish the populations of what Edward O. Wilson describes in *The Diversity of Life* as "the large, the slow, and the tasty."[1] But for the dawn of agriculture many millennia ago, our species might have remained a bit player in the continuous network of give and take, a small but enduring part of the flowing interactions among animals, plants, and microbes. As it is, the bending of natural systems in ways that favor our prosperity has become fundamental to human cultures worldwide. The underlying premise of civilization, after all, is the transformation of wildness to something "civil." Missing from the process, perhaps until now, is a recognition of our fundamental dependence on the wildness and on the small ingredients—call them species, if you will—that make natural systems what they are and cause them to function as they do.

Late in the twentieth century, the importance of the ocean to all earthlings became increasingly clear as the phenomenon known as El Niño, a warm shift in a normally cold current off the coast of Peru, triggered events half a world away—floods, droughts, unusually high and low temperatures, outbreaks of disease, changes in crops, certain ups and downs in the stock market, and much more. Headlines around the world seemed to indicate that people might be awakening to the links between the ocean and their everyday lives and, perhaps, recognizing as never before that the sea is the driving force behind climate, weather, thermoregulation, and planetary chemistry. Still missing from general awareness, however, are the vital importance of the sea as the cornerstone of Earth's life-support system and the fact that, in large measure, the sea is what it is because of the life it contains.

A thousand years ago, well into the long warm phase that enabled humankind to develop agriculture, wildlife abounded, both on land and in the sea. Ever since then, this abundance and diversity have steadily declined, coincident with a rise in the number of people. Slowly at first, but with accelerating momentum, losses have been piling up—of species, then of entire ecosystems, and, over time, of the underlying natural systems and processes that support us all. Loss of what we think of as "wild" translates to consumption of our life-support system, the steady and deadly erosion of the foundation on which civilization rests.

When Leif Eriksson explored the North Atlantic Ocean at the beginning of this millennium, the continents were lightly populated, by only a few hundred million of us worldwide. As the great era of discovery unfolded in the thirteenth and fourteenth centuries, natural systems gave way as human cultures prospered. Yet as recently as some five hundred years ago, when Ferdinand Magellan first circumnavigated the world and Vasco da Gama pioneered a sea route linking Europe and India, much of the land and essentially all of the sea was as pristine as it had been for thousands of years. This held true even into the eighteenth century, when Captain James Cook journeyed around the globe, and continued with little change until about two centuries ago, when Lord Byron wrote these words:

> Roll on, thou deep and dark blue ocean—roll!
> . . . .
> Man marks the earth with ruin—his control
> Stops with the shore.

As the nineteenth century began and the human population reached 1 billion, the pace of change picked up. Meriwether Lewis and William Clark opened the way for development of the American West while whalers, sealers, and fishermen extended the principles of hunting and gathering wildlife—farther, deeper, longer—into unknown oceans. In the twentieth century, technology advanced swiftly, enabling those who would conquer wilderness to be increasingly effective in doing so, both on land and in the sea. Coastal areas, where more than half of the world's people choose to live, were heavily affected, partly because fishing pressure was greatest near the most populous areas but also through pollution and direct shoreline actions—dredging, filling, carving of channels, walling of beaches, and elimination of marshes, coastal forests, and dunes. Although numbers matter, even a handful of individuals armed with bulldozers or bombs or chemicals or trawls proved capable of undoing in minutes systems that had been honed over millions of years.

Today, with the world's population nearly six times what it was in Lord Byron's time, our ruinous ways have breached the shore and extended to the deepest ocean. Many people still stubbornly cling to the notion that humankind is too puny to alter the nature of the sea, yet simple mathematics tell us that increasing numbers of people mean increasing consumption of natural resources and increasing amounts of pollution. Some pollutants are visible: plastic debris, lost fishing gear, bottles, cans, and other trash. Others are less obvious but can be deadly when allowed to flow into the sea via groundwater, rivers, and streams: pesticides, heavy metals, toxic chemicals, and nitrates and phosphates from excessive doses of fertilizer applied to lawns, fields, farms, and golf courses. The ocean's changed chemistry affects the biology of life, sometimes by altering a creature's reproductive potential, sometimes by reducing its resilience, sometimes by causing outright death. There is no doubt that wildlife in the ocean is under apocalyptic siege by humankind, not only through the noxious substances we allow to flow into the sea but also, even more strikingly, through what we take out.

In the 1960s, the nations of the world embarked on a deliberate and ambitious campaign to extract as much as 100 million tons of wild-caught fish, shrimp, and other ocean wildlife from the sea annually, an amount that some thought reasonable and sustainable. This required fishing fleets to be ramped up, new technologies to be adopted, and government subsidies to be accepted. Had we declared war on tuna, cod, swordfish, herring, hake, halibut, oysters, clams, lobsters, and more than a hundred other commercially targeted species, the effect could not have been more deadly.

In 1991, while serving as chief scientist of the National Oceanic and Atmospheric Administration, which includes the National Marine Fisheries Service as one of five divisions, I was stunned by a report I received that documented the decline in numbers of Atlantic bluefin tuna to 10 percent of what they had been twenty years before. At a meeting held to discuss the problem, I blurted out, "What are we trying to do? Exterminate them? If so, good job! We're ninety percent there."

Unfortunately, bluefin tuna are not the only ocean species now in a precarious state as a consequence of aggressive predation on our part. Humans, heretofore a largely terrestrial species, are now moving into the sea armed with lines as much as seventy miles long containing clusters of baited hooks every few feet, and with nearly invisible nets of new lightweight, durable materials stretched over thousands of miles of ocean, indiscriminately catching and killing whatever becomes entangled. Aircraft are used to spot concentrations of fish, birds, or squid and then to vector in boats equipped with acoustic tracking devices and nets to encircle and capture every last member of the populations found.

Over the past twenty-five years, the take of ocean wildlife first increased, reaching a maximum of about ninety tons in 1989, and then declined, despite greatly enlarged fishing fleets and the use of ever more sophisticated acoustic fish-finding devices, pinpoint navigation, and other new technologies. By 1996, the evidence of excessive exploitation was compelling. A report from the World Conservation Union documented what many had come to fear: more than 100 marine species were shown to be in trouble and were listed as either threatened or endangered. The list included species that are recognized and for which some documentation exists. But what of the others about which little or nothing is known?

Even on land, the magnitude of our ignorance about the extent of biodiversity loss hampers the development of appropriate policies and action plans. Simple questions such as "How many species of organisms are there?" have a range of answers, from the number so far described (plus or minus 1.5 million) to the number likely to be out there (unnamed, but educated estimates range from 5 million to 120 million).

The sea contains more high-level taxonomic diversity than do terrestrial environments—by a wide margin. Of the thirty-three or so phyla of animals, thirty-two live in the sea; twelve occur on land. Ninety percent of all known classes are marine. But to appreciate fully the diversity and abundance of life in the sea, it helps to think small. Every spoonful of ocean water contains life, on the order of 10^2 to 10^6 bacterial cells per cubic centimeter plus assorted microscopic plants and animals, including larvae of organisms ranging from sponges and corals to starfish and clams and much, much more.

In effect, the entire liquid mantle embracing our planet is a living minestrone, with the majority of the bits small or microscopic. Most people ignore microbeasts, yet as microbiologists are fond of pointing out, most of the biochemical action that shapes the biological and much of the physical and chemical character of Earth is accomplished by microbes. Physiologically, these small creatures, both above the ocean's surface and below, are more diverse than are plants and animals combined. With a broad spectrum of capabilities and lifestyles, they range from free-living, photosynthetic, and chemosynthetic autotrophs to creatures that live on and decompose all naturally produced organic materials.

In 1997, a new kingdom of life-forms was officially recognized, and not surprisingly, it is microbial and marine. The Archaea had been discovered twenty years earlier in association with hydrothermal vents in the deep sea. The awareness that they exist, and probably have existed for more than a billion years, should serve as a wake-up call for us—a call to explore the ocean. After all, less than 5 percent of it has been explored and mapped with the same level of detail accorded the moon and Mars.

What discoveries await us? Based on what we already know, the most frightening news may be the magnitude of what we have yet to grasp about the nature of the living world that supports us and what we need to learn in order to maintain our cultures, our civilization, within this natural framework. We need to know, but are far from understanding, how to use the species and systems that surround us without using them up.

Botanist and biodiversity expert Peter Raven suggests that the world has not been managed sustainably for humankind since World War II. He gives as evidence the loss of 25 percent of the world's topsoil since that time; large changes in the composition of the atmosphere; the loss, without replanting, of about one-third of the world's forests; the loss of 15 to 20 percent of the world's arable land; and threats to biodiversity that may drive to extinction as many as one-fourth of the world's species of plants, animals, and other organisms within the next twenty-five to thirty years. In a 1995 paper titled "The Importance of Biodiversity," Raven pointed out:

> Assuming a current extinction rate of five per cent of species per decade over the next thirty years, accelerating with the addition of 90 million people a year to the world population, we shall probably lose, very approximately, something like 50,000 species a year over the next several decades, of which only something like 7000 (about a seventh) would have ever been known or recognized before we lose them.[2]

Raven's estimates are based largely on what is known of terrestrial systems. When the unknown, unexplored ocean systems are taken into account, the status of future biodiversity may be even more grim. Marjorie Reaka-Kudla, in *Biodiversity II,* comments on just one of the ocean's more obvious trouble spots:

> Coral reefs may contain far more species than previously supposed . . . and very large amounts of this biodiversity may be lost due to human activities before they are even discovered.[3]

I thought of such things recently as I immersed myself in a warm, dark ocean, brushed by soft, gelatinous jellies; wondered about the fate of a school of silver jacks sweeping by, catching the light like bits of broken glass; watched as a thumb-sized squid flashed rainbow colors and paused to look me over before jetting off, perhaps thinking its own molluskan thoughts. I mused about the consequences of events in 1998, designated by the United Nations and the United States government as the Year of the Ocean. Many people from many nations gathered in Lisbon to celebrate—and deliberate—the past, the present, and the future of humankind's relationship with the sea. Speeches were made; there were parades and fireworks. In Monterey, California, the president, vice president, and first lady of the United States joined with more than 500 invited guests to focus on ocean issues and begin developing a plan of action for the next century.

As I write, optimism is running high that the ocean may become a priority for attention in the twenty-first century, beyond what space has been in the twentieth: a focal point for exploration and technology development, but also for establishment of policies to ensure sustained, peaceful use rather than runaway exploitation, in an arena of conflict. On this subject, Boyce Thorne-Miller is at her best, offering insight and clear thinking about conservation of the living ocean, right down to practical matters involving laws, protected areas, protective policies, and an underlying ethic of caring that is at the heart of what this book is all about. Of all the threats to the future of the ocean, to life in the ocean, and thus to our own future, lack of understanding tops the list. As an antidote, reading *The Living Ocean* is a fine place to begin.

Sylvia A. Earle
Explorer in Residence, National Geographic Society
Spokesperson, SeaWeb

Acknowledgments

I am grateful to Friends of the Earth U.S. for supporting the publication of the first edition of this book and for making it possible for me to do this second edition under the auspices of SeaWeb. I also thank Friends of the Earth International for having me represent them in several international forums involving ocean conservation. That experience has given me insight that would otherwise have been unattainable.

I owe a very large thank-you to Vikki Spruill, executive director of SeaWeb, for her support and patience while I wrote this second edition, which grew into far more than anticipated, and for her friendship along the way. I also thank Pew Charitable Trusts for supporting this effort and everyone at SeaWeb for their help. A very special thanks to Bruce McKay for reviewing the draft manuscript, for providing many good suggestions and inspiring discussions, and for steering me to important literature.

I am forever grateful to Clif Curtis for ten years of mentoring. He made the first edition of this book possible, and he has continued to provide guidance, information, and opportunities to work with national and international networks of environmental organizations. His contributions are reflected throughout this text. I thank John Catena for his input as coauthor of the first edition, without which this new text would never have been born. Sally Ann Lentz, Beth Millemann, and Kieran Mulvaney have also contributed immeasurably to my understanding of marine affairs, and I have cherished their camaraderie as we all try to "save" the ocean. Richard Ballantine has been a mentor, a friend, and a fellow ocean enthusiast through the trials and tribulations of my limited publishing career.

Jackson Davis, Paul Dayton, Richard Gammon, Kristina Gjerde, Fred Grassle, Jack Hardy, Paul Johnston, John McGowan, Laury Miller, Mike Mullin, Ted Smayda, Peter Taylor, and Michael Weber have been invaluable sources of information for this book. Thanks to all these people and to numerous others who have shared information and insights pertinent to the topics discussed herein. And thanks to my family and friends for putting up with me throughout this endeavor.

Finally, my heartfelt thanks and endless admiration to Sylvia Earle for her moral support, the knowledge she has shared, her love of the ocean, and her belief in the importance of the vast variety of life in the sea. She is one of a very few human beings who visit marine creatures in their own home and experience them on their own terms as she explores some of the ocean's magnificent depth and breadth by scuba and submersible. Her scientific and personal impressions of that realm give credence to my belief that precaution and protection are essential for the entire interactive network of ecosystems that make up the world's ocean.

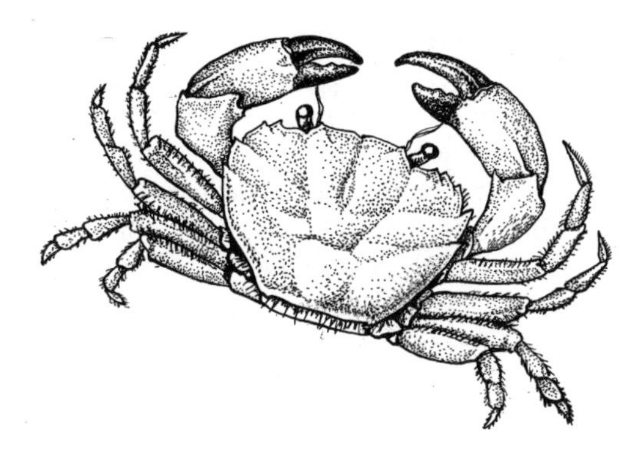

Introduction

If life on earth has a single outstanding property, it is that it exists in an enormous variety of forms. This was clearly evident to Charles Darwin during the famous voyage that led to his theory of evolution to explain the seemingly unending diversity of life-forms that he observed. It is even more evident to those evolutionists who have studied the fossil record and proposed new theories that modify or counter Darwinism. It is sobering to read that more than 99.99 percent of all the species that have ever lived on earth are now extinct.[1] Nevertheless, the array of extant species is enough to amaze anyone who travels the natural areas of this planet, including those who explore the living world within the ocean. Yet despite this obvious and scientifically tantalizing variety, only a fraction of the species living today are known to science, and many will never be known as living species because they will disappear before they are discovered. This is by now a familiar story for the tropical rain forest, and it is being told more often for coral reefs, but the story of biological diversity in the rest of the ocean is all but unheard.

As human beings have populated the lands of the earth, we have pushed out other forms of life. For a time, it seemed to some that our negative influence must stop at the ocean's edge, but that has not proved to be so. By overfishing the living bounty of the seas and by flushing the wastes and by-products of our societies into the oceans, we have managed to impoverish, if not destroy, living systems there as well. The world's ocean covers 70 percent of the earth's surface and, when depth is considered, contains nearly one hundred times more inhabited space than the continents.

Unfortunately, so little is known about the variety and distribution of ocean species and about the living processes characteristic of marine ecosystems that it is not yet possible to fully assess the losses, either qualitatively or quantitatively. We learn more each day, but each day we also lose forever the chance of learning about some marine animals and systems.

One thing we do know is that there is a broader spectrum of life-forms in the ocean than on land. This is reflected not in the number of species but in the numbers of higher taxa (families, orders, and phyla), which represent greater genetic differences than occur at the species level. If there are more species on land than in the ocean, as is currently thought to be true, it means only that there are more closely related species on land. How should we weigh the importance of small genetic variations against the importance of very large genetic variations among species?

Furthermore, the list of existing marine species has not yet been fully compiled. Certainly, the number known to science is rapidly growing, and it appears that early estimates are far too low. For instance, scientists are discovering that the deep-ocean floor, originally thought to be biologically poor, supports a diversity of species that may rival that of the tropical rain forests. Almost nothing is known of the numbers and species distribution of many microbial organisms that live in the ocean, and recent research has revealed that certain life-forms previously described as single species on the basis of form are actually several species on the basis of molecular genetics. Moreover, rare species may not be as uncommon in the marine environment as has been assumed; it is more likely that many simply have not been noticed or identified because marine ecosystems are so vast, varied, and unexplored compared with terrestrial ecosystems. Species may disappear within the sea without a ripple being observed by the human eye.

It is easier to estimate the damage done by habitat destruction in most terrestrial environments than it is to estimate the damage caused by alterations in the biological and chemical dynamics of underwater ecosystems. The genetic variety and the vitality of marine ecosystems are suppressed by toxic chemical pollution, eutrophication leading to anaerobic conditions, and overfishing. Such stresses lead to biological impoverishment, in which species may not disappear altogether but their population and genetic variations are reduced. Impoverished ecosystems can readily be pushed to the brink of collapse. In such an unhealthy state, relatively small environmental stresses may trigger widespread biological losses, including extinction of species, at least on a local scale. Because threats to marine biodiversity are difficult to quantify, they are often simply overlooked.

A major problem in assessing both marine and terrestrial biodiversity is the paucity of taxonomists and systematists—scientists who study the identities of living organisms and the genetic relationships among them, respectively. Another is the unwillingness of governments to fund research that does not have immediate application to policy and regulatory decisions.

The recent discovery of an entirely new group of organisms, the prochlorophytes, previously overlooked yet now known to account for a significant proportion of primary production in the sea, merely emphasizes how much there still is to learn about life on earth, especially in the marine environment. Yet basic descriptive research in the ocean is prohibitively expensive, thus requiring government support and citizen interest.

In addition to knowing what forms of life and what species live in the sea, it is important to understand how they live. Recent global views of the "living planet" have turned attention to the importance of the variety of biological functions performed by the various species, since these functions help maintain the geochemical cycles that make the earth hospitable to life as we know it. On a smaller scale, species are principal functional units of ecosystems, and the health of a given ecosystem depends on its living members being able to perform their roles well. Until we can better identify marine species and their functions, we will not be in a strong position to effectively regulate human activities that are potentially harmful to important elements of the biosphere. In the meantime, it would be prudent for human societies to regulate themselves by taking a precautionary approach: if there is a reasonable likelihood that an activity will cause negative environmental effects, even in the absence of scientific proof, action should be taken to eliminate the potential for damage to the ecosystem.

The relative importance of ocean ecosystems in maintaining the conditions that support life on earth have only crudely been estimated and modeled. However, as terrestrial ecosystems become ever more depleted by human expansion and development, the importance of the ocean in maintaining life-supporting geochemical cycles only increases. It would be a mistake to underestimate our reliance on a healthy global ocean. The international political community is beginning to recognize that, and several important initiatives to protect the ocean environment have been agreed on and now need to be implemented by the nations of the world.

This book is intended to be a brief introduction to the science and policy of biodiversity in marine environments and the relationship of human societies and governments to the living ocean and the need to protect it. It includes an overview of basic concepts and principles that apply to biodiversity in marine environments and reviews the policy issues and existing instruments pertinent to the protection of that biodiversity. Since the first edition of this book was published, scientists have learned a great deal more about marine biodiversity and its decline. There have also been important new international measures addressing threats to marine biodiversity. Consequently, this second edition cannot adequately summarize all the available information and merely introduces the subject, with the hope of inspiring readers to become concerned and to pursue more detailed information among the references listed in the bibliography. With 1998 designated by the United Nations as the Year of the Ocean, this is an appropriate time to raise people's consciousness about the biodiversity of the ocean and to suggest a path toward implementing increased protection for living marine ecosystems.

Biodiversity is defined in Chapter 1, and the importance of threats to marine biodiversity are discussed in Chapter 2. A brief review of the current status of scientific knowledge with respect to marine biodiversity is presented in Chapter 3. Chapters 4 and 5 describe the major coastal and oceanic ecosystem types and comment on the major threats to biodiversity in each. This is followed, in Chapter 6, by a general discussion of the ways in which government and the public can protect marine biodiversity; Chapter 7 contains specific examples of national and international policies, legal instruments, programs, and institutions. The final chapter is a statement about attitudes and approaches to protecting marine biodiversity. The bibliography includes all references used in the preparation of this book, whether or not they are cited in the chapter endnotes. The endnotes themselves have, in most cases, been cited at the ends of paragraphs so as not to interrupt the text; thus, the references listed in an endnote refer to various parts of the paragraph, not specifically to the ending sentence.

Finally, considering the rapidly growing and evolving body of scientific information, policy and legal frameworks, and management methodology, it is essential that students of policy and policy makers themselves consult the experts who understand the complexities of the ecosystems, the legal systems, and the social systems involved. Environmental conservation is an interdisciplinary field in which science, economics, law, sociology, politics, ethics, and religion intermingle. If effective environmental policies—be they marine or terrestrial—are to be crafted, experts from the various disciplines must communicate with one another and with members of the general public, who in the end will be the ones to implement or approve implementation of the protection of biodiversity on our planet.

Chapter One

The Importance of Marine Biodiversity

It is difficult to assess the number of species in the marine environment or their abundance, status (whether their populations are increasing or decreasing), and genetic variation. Aside from seabirds, marine mammals (e.g., whales, seals, sea otters), and sea turtles, which are among the most obvious and charismatic species as well as those commonly targeted for slaughter in the past, endangered or threatened status has been assigned to few marine species. It has generally been thought that ocean species are dispersed widely and have great reproductive capacity, and as a consequence the decline or disappearance of a given species in one place would not threaten the species globally. However, there are many marine species for which this is not true, and the potency and magnitude of human impacts on the ocean environment are now recognized to be far greater than previously suspected. Furthermore, it is becoming ever more apparent that the diversity of species in the ocean has been grossly underestimated, especially in remote environments such as the deep-ocean floor.

It is important to understand that the absence of knowledge about endangered species or recent species extinctions is different from knowledge about the absence of such occurrences. Studies of the deep-sea floor and even shallower waters have revealed numerous unknown species, amounting to 50 to 80 percent of species in each sample, living in the sediments. Coral reefs are the marine ecosystems most ostentatiously rich in species, and even though they are among the most visited and best-known marine areas, they still harbor numerous unknown species. If so many species remain unknown, unstudied, and untallied, it is impossible to know how many are confined to relatively small geographic areas and which ones are indeed threatened or endangered.[1]

Modern extinctions in the ocean may not be as rare as once thought. A few species have turned up missing, and that has led marine biologists to suspect that others may have disappeared unnoticed. At one extreme are estimates that at least 1,200 marine species have become extinct in the past few hundred years and that many thousands or more than a million could disappear in the next hundred. Certainly, few extinctions have been documented, and those have generally been among the larger, conspicuous marine animals with relatively small populations, such as marine mammals. Only recently have marine biologists documented species extinctions among the less conspicuous but far more common sea life, such as invertebrates.[2]

Whether these extinctions are symbolic of others that have occurred but have not been documented by science or whether they are precursors of future extinctions that will occur as marine ecosystems become more physically and chemically stressed and biologically depleted, it is clear that the potential for marine extinctions has been vastly underestimated in the past. And perhaps species extinctions are not the most important story in the ocean—maybe depleted populations and diminished diversity within ecosystems (degradation of the structure of biotic communities) are equally threatening.

What Is Biodiversity?

The term *biodiversity* (or *biological diversity*) has more than one interpretation, so it is important to understand the context in which the term is used. The purist would use *biodiversity* to refer only to the number and identification of species present in a particular ecological system. The species is the common unit of evolution, so diversity at the species level has special importance. However, biodiversity also may be assessed at higher levels of taxonomic classification of animals and plants (genus, family, order, and phylum, in ascending rank). Diversity at these levels is particularly important in the marine environment.

Recognizing that taxonomic classification is not the only way of categorizing the biological world, scientists sometimes view biodiversity in other terms as well. Most scientists now recognize the following categories of biodiversity:[3]

- *Species diversity* or *taxonomic diversity,* which refers to the variety of species or other taxonomic groups in an ecosystem
- *Ecological diversity,* which refers to the variety of types of biological communities found on earth
- *Genetic diversity,* which refers to the genetic variation that occurs among members of the same species

- *Functional diversity,* which refers to the variety of biological processes, or functions, characteristic of a particular ecosystem

When comparing the species diversity of different ecosystems, it is important to consider, if possible, the *characteristic diversity* of each system—that is, the species diversity typical of the unstressed or normally stressed ecosystem of a particular type. There are significant differences among ecosystems and geographic locations. For example, natural (unchanged by human intervention) polar ecosystems characteristically have fewer species than do natural tropical ecosystems. Yet it would be a mistake to say that tropical ecosystems are therefore more important. What is important is that natural ecosystems have greater species diversity than do abnormally stressed versions of the same type of ecosystem. When selecting refuges of biodiversity for special protection, it is important to consider ecological diversity and identify representative "hot spots" of species diversity for various ecosystem types.[4]

Functional diversity may or may not mirror species diversity, but it does reflect the biological complexity of an ecosystem. Some scientists argue that examining functional diversity may in fact be the most meaningful way of assessing biodiversity in the oceans while avoiding the quagmire of cataloging all species, a difficult and usually impossible task in marine ecosystems. By focusing on process, it may be easier to determine how an ecosystem can most effectively be protected. Interfering with an ecosystem function may be more readily understood as harmful to the greater ecosystem, and therefore more important to humans, than endangering a species for which nobody has yet found a use (it is an unfortunate truth that many people still value nonhuman life only in terms of its immediate usefulness to humans). After all, protecting biological functions will protect many of the species that perform them. Nevertheless, it is difficult to identify all the important functions operating in a complex ecosystem. Detecting some functions may, in the end, boil down to identifying and studying the species that perform them.

There are opposing theories about the association between functions and species and whether species are unique in their particular functions. On one side, some scientists suggest that as long as the functions of an ecosystem are being carried out, it doesn't matter which or how many species are performing them. This philosophy would suggest that the loss of species from an ecosystem is not alarming if other species move in to take their place, a situation that can happen rapidly in marine ecosystems. These scientists believe that there is unessential redundancy in ecosystems.[5]

On the other side, some scientists have noted that the greater is an ecosystem's species diversity, the more functions the ecosystem is able to carry out and the more efficiently it does so; moreover, the more species present that have similar functions, the more stable the ecosystem. These scientists believe that there is no redundancy or, if there is, it adds to the ability of a system to survive, adapt, and evolve.[6] Scientific evidence is mounting in support of this viewpoint. However, it may be more applicable on short time scales than on long ones. From the perspective of the earth's biosphere, it may not matter whether ecosystems topple and are eventually replaced with new ecosystems containing new complements of species. But whether humans would appreciate the change or still be part of the scene is unlikely. Thus, it seems that we would have a vested interest in protecting, for as long as possible, the full complement of species associated with our own and stabilizing the ecosystems on which we rely.

The Importance of Biodiversity

In the face of environmental change, loss of genetic diversity weakens the ability of a species to adapt; loss of species diversity weakens the ability of a biological community to adapt; loss of functional diversity weakens the ability of an ecosystem to adapt; and loss of ecological diversity weakens the ability of the entire biosphere to adapt. Because biological and physical processes are interactive, losses of biodiversity may precipitate further environmental change, which may prevent the recovery of lost diversity and may lead to further losses. This progressively destructive routine results in biologically impoverished ecosystems, in which species are fewer and/or genetic variation is severely reduced.[7]

An impoverished ecosystem is more susceptible to failure (loss of capacity to function as a complex, productive system) when faced with additional environmental changes, be they natural or of human cause. Just when the final straw will fall and what it will be for any particular ecosystem is not easily predictable. Consequently, we can be lulled into the false conviction that particular ecosystems and even the biosphere as a whole are not in imminent danger because systems seem to be functioning well with their significantly reduced biodiversity. Yet they may be on the brink of failure—certainly some, such as the North Atlantic coastal ecosystems that once supported rich fisheries, have already crashed. The destruction can come swiftly and offer no opportunity for recovery—just a long wait for a new system, somewhat different in nature, to establish itself.

There are two general categories of reasons to protect biodiversity, one based on arguments of principles, ethics, and spirituality and the other based on human self-interest. In the first way of thinking, other species have a right to exist and have a place in the global ecosystem, and therefore it is simply wrong to drive them to extinction. Furthermore, ecosystems and their species are regarded by religions as the Deity's creation. Their destruction diminishes the Creation and should be avoided; some religious leaders go so far as to call it a sin. The result of this diminishment may be the impoverishment of the human spirit as well as of Creation. The other justification for protecting species is that they are of direct or indirect value to humans, either as commercial resources or as providers of valuable services. Whereas the ethical reasons apply under all circumstances, the human-centered reasons apply only to those species or ecosystems for which there is a perceived human value now or in the future. Some species are not deemed valuable; some species and services are unknown and therefore overlooked; and future importance is difficult to predict. Human-centered considerations merge with ethical considerations when it is argued that biodiversity should be protected for the potential benefit of future generations.[8]

Biodiversity is important on global, ecological, and human scales. Biological processes regulate global climate and cycling of essential elements and substances and as a result perpetuate a global system favorable to the support of life. Greater species diversity seems to contribute greater stability to ecosystems, and conversely, the healthy functioning of a variety of ecosystems supports a greater diversity of life. From the human perspective, our societies need living resources, and greater biodiversity offers a greater variety of foods, building materials, medicines, and products of various sorts. On a more intimate level, a person may appreciate biodiversity and living organisms for aesthetic, psychological, and religious or spiritual reasons.

It has been amply demonstrated that many human activities have placed stress on marine ecosystems, resulting in the biological impoverishment of coastal ecosystems and possibly threatening open-ocean ecosystems more than we have realized in the past. The consequences of this impoverishment are likely to be far-reaching. Yet human populations and activities continue to threaten biodiversity on earth. As humans, we are alone in being able to destroy vast numbers of species and understand what is happening, and we are alone in being able to end this destruction by our own actions rather than by our own extinction.

The Global Significance of the Ocean and Its Life

James E. Lovelock, in his Gaia hypothesis, views the earth, with its biota, rocks, air, and oceans, as a single, intercoordinated entity. He suggests that the earth's environment has coevolved with its biota, resulting in self-regulation of the global ecosystem's climate and chemistry. Although much of the theory is controversial, and it is better applied on geologic time scales, it provides a useful metaphor for considering interactions between the earth's environment and biota in the shorter term.[9]

Numerous apparent feedback mechanisms between living and nonliving components of the earth make the Gaia hypothesis enticing—the potential role of forests, especially tropical rain forests, in regulating global warming, for instance. Forests move vast volumes of water from the soil to the atmosphere, and the rate of this process increases in response to warming of the atmosphere. As global warming takes place, the increased moisture in the warmer air is predicted to give rise to more clouds, which will shade the earth, reducing temperatures and thus moderating the warming effect. Trees also take up carbon dioxide and thus moderate its rate of increase in the atmosphere due to human activities. It is even possible that increases in carbon dioxide will stimulate higher rates of photosynthesis and therefore higher rates of uptake of the gases that contribute to global warming—if there are any trees left. But trees are not the only photosynthesizers capable of changing the concentrations of atmospheric gases. The microscopic plants of the ocean are a major source of oxygen and an important sink for carbon dioxide.

Several global climate and chemical feedback mechanisms have been proposed for the marine ecosystem. It was long speculated that volatile sulfur compounds produced by marine organisms might explain the measurable movement of sulfur from the sea to the land. Indeed, it has been verified that certain species of microscopic algae in the sea surface waters produce such a compound—dimethyl sulfide (DMS)—in large quantities. When released into the atmosphere, this compound oxidizes to form a sulfate aerosol, which is recognized as the major source of condensation nuclei for cloud formation over the ocean. Furthermore, it has been suggested that production of this compound will increase as average global temperatures rise and warmer ocean surface waters favor growth of the species that produce DMS; the resulting increased cloud cover would have a cooling effect due to the backscatter of incoming radiation by the clouds.[10]

The ocean is widely recognized for its important role in the carbon cycle and therefore in adjusting the concentration of carbon dioxide in the atmosphere. Hundreds of different species of microscopic plants, known as *microalgae* or *phytoplankton,* grow in the zone of light, which varies from about 10 to 200 meters (33–660 feet) at the surface of the ocean. They use carbon dioxide from the water as they photosynthesize, creating a deficit of carbon dioxide in the surface waters. Carbon dioxide is added to surface waters by three processes: absorption of carbon dioxide from the atmosphere, release of carbon dioxide from the respiration of bacteria and animals in the water, and upwelling of carbon dioxide from dissolved calcium carbonate in deposits of shells and other materials on the ocean floor. The way in which these processes balance out varies with time and location in the ocean and may result in the net absorption of carbon dioxide from the atmosphere or the net release of carbon dioxide into the atmosphere.[11]

The role of phytoplankton in the carbon cycle has led scientists to postulate about the potential for these species to regulate atmospheric carbon dioxide and abate global warming. It is difficult to predict whether an increase in atmospheric carbon dioxide and/or a rise in temperature of the ocean surface waters will promote an increase in phytoplankton photosynthesis, thereby creating a feedback mechanism for reducing the excess carbon dioxide in the atmosphere (generated by industrial activities) and thereby moderating the warming effects. One intriguing prediction is that the warmer waters will promote the growth of a particular type of phytoplankton called coccolithophores, which produce shell-like cell walls made of calcium carbonate and therefore have a greater requirement for carbon dioxide. If that were to happen, some of the excess carbon dioxide would indeed be removed from the atmosphere.

Because of the real threats posed by climate change, various "climate engineering" schemes have been proposed. One such proposal, known as the "Geritol solution" to global warming, involves human intervention in natural systems. Large portions of ocean surface waters are not very productive because of limited amounts of iron, a nutrient required by phytoplankton, so it was suggested that a solution of iron be spread over great areas of the ocean to promote phytoplankton growth and photosynthesis, thus causing the uptake of carbon dioxide from the atmosphere. Two tests in the eastern Pacific Ocean revealed that algae did "bloom" when supplied with iron, as long as fresh iron was added every couple of days, since it quickly precipitated and sank. Many concerns have been aired, however, because such tampering, even if successful, could have unpredictable and unwanted effects on the ecosystem's food chain and biodiversity. Other scientists doubt that it would work or that it would be feasible to introduce as much iron as would be needed.[12]

Phytoplankton also produce oxygen in the process of photosynthesis. An estimated 50–75 percent of the atmosphere's oxygen comes from ocean photosynthesis. Moreover, phytoplankton plays a crucial role in the movement of nitrogen, phosphorus, silicon, iron, and other elements through the earth's environments. The abundance and species diversity of these tiny plants can be expected to have a pronounced effect on these critical cycles.[13]

The ocean's fluidity and constant motion facilitate interconnectedness among its ecosystems and with the land and the atmosphere. It has been claimed that the ocean is the engine that drives the earth's living system. The ocean and its life regulate climate and make the earth habitable; it is the kingpin in the major cycles of important chemicals of life; it is the source of water that makes life on land possible; and it is home to most of life on earth. Ocean currents move water downward, upward, and about the globe. Water and the dissolved materials and organisms floating in it move freely from one ecosystem to another, with many species living different parts of their life cycles in different ecosystems. What happens in one part of the ocean is often eventually felt in other, far-removed locations.

There is further evidence of the interaction between ocean biota and global climate. The geologic record in ocean sediments and polar ice sheets suggests that variations in both carbon dioxide level and climate are linked in some way to changes in the amount of carbon taken from the atmosphere and laid down, as organic matter or shells, in marine sediments. It is not known just how climate change and change in the ocean carbon cycle are related, but scientists have speculated that increased productivity of oceans has caused global cooling by reducing atmospheric carbon dioxide in the past.[14]

Benefits of Marine Biodiversity for Humans

Numerous marine species can be considered "goods" for human use, consumption, and commerce. Living marine resources may be a source of food and food supplements; compounds used in drugs and for other medical purposes; cosmetics; brood stock for aquaculture; animal feed or fertilizer in aquaculture, agriculture, and pet food; genetic stock for aquaculture and research; genes for biotechnology applications; materials for use in food processing and other industrial applications; building materials; live aquarium animals; ornamental objects; fur coats; and objects and substances for use in traditional healing practices and alternative medicine.

In addition to feeding people, fishing provides a livelihood for coastal and high-seas fishers around the world. The importance of food from the sea varies considerably from nation to nation. In some countries, it is one of the primary sources of protein. The populations of some thirty-eight countries rely on fish for more than 30 percent of their animal protein intake and more than 10 percent of their total protein. Six of the top eleven fishing nations (China, Chile, Peru, India, the Republic of Korea, and Indonesia) are developing countries. As populations grow and fishery yields decline, aquaculture is being rapidly developed to supplement the global production of seafood.[15]

An estimated 9,000 species of fish are currently exploited, but only 22 of these are harvested in global-scale quantities (exceeding 100,000 metric tons, or 110,160 short tons, annually, for export markets). Of these, the seventeen major oceanic fisheries are all being fished at or beyond capacity and nine are in a state of decline, among them herring, cod, jack, redfish, and mackerel, which in the 1980s accounted for about half the annual world catch. The noncommercial taking of seaweeds, shellfish, and fish for food by coastal peoples around the world is also declining precipitously.[16]

During the 1950s and 1960s, there was a great deal of public speculation about how the oceans would solve future world food problems. It was predicted that new food sources would be exploited and large-scale sea farming would become popular. This prophecy has not come true, and such speculation is no longer popular. Most fisheries have reached or surpassed their sustainable yield, and many are severely depressed. A few species formerly discarded as "trash fish" have been marketed, but most go into the production of fish meal for pet food and aquaculture and the feeding of animals that should be eating plants—chickens, cows, and pigs. Sea plants have not proved to be a major food worldwide, as once envisioned, but they continue to be harvested and marketed in cultures that have historically used them, such as Japan, and in health food and gourmet food outlets.

It is not uncommon for marine organisms to produce toxins that repel predators or retard the growth of competing species. Such chemicals that are not used directly in the metabolism of the organism are known as *secondary metabolites.* A direct consequence of this "chemical warfare" in the sea is the production of a considerable variety of biological substances that have medicinal value for humans. Derivatives from marine flora and fauna include numerous drugs with various properties: antibiotics (both antimicrobial and antiviral), tumor inhibitors, coagulants and anticoagulants, drugs for treating heart and nerve ailments, pain suppressants, anti-inflammatory agents, skin care products, and sunscreens.[17]

Such pharmaceuticals are derived from many marine organisms, among them seaweeds and invertebrates from coral reef habitats, where the production of defensive chemicals is both commonplace and effective. Several algae, corals, sea anemones, sponges, and mollusks have antibiotic and/or anticarcinogenic properties that could be isolated for medicines. Coral is used in the reconstruction of damaged bones, and oyster shells are a primary source of calcium supplements. The porcupine fish and the puffer fish have yielded symptomatic treatments for terminal cancer, and shark's liver contains substances that enhance resistance to cancer. Sea cucumbers, sea snakes, menhaden, and stingrays produce materials useful in the treatment of an array of cardiovascular ailments, and extracts from seaweeds and octopuses treat hypertension. Substances isolated from seaweeds are active against viruses causing cold sores, eye infections, and venereal disease, and a sponge has yielded a substance effective against viral encephalitis. And the sea may lead the way to the successful treatment of viruses such as the common cold.[18]

The advent of genetic engineering provides an opportunity to enhance production of marine metabolites that are useful in medicine. Often, the need for important genes for biotechnology is used as a justification for preserving genetic and species diversity. As one scientist put it, biotechnology offers a "strong foundation for exploitation of biologically active compounds already known to occur in the sea and for further exploration into the recesses of the world oceans for compounds and food sources as yet undiscovered." In fact, the unusual deep-sea communities clustered around "hot vents" on the sea floor offer a wealth of different microorganisms with promise for biotechnological uses. In marine aquaculture, biotechnology is also being used to produce certain desirable attributes in the cultured animals or plants, as in the case of fish being modified to grow bigger and faster, and genetic alteration has been suggested for the production of strains that are incapable of reproducing if they escape into the natural environment. The techniques used in biotechnology have also provided important information about genetic biodiversity in the ocean. However, this research can be accomplished without the exploitation of species to produce new genetic stock, so these advances in genetic research should not be attributed to biotechnology.[19]

There are concerns as well as advantages associated with the production of new organisms through biotechnology; the real dangers of genetic manipulation, of which there are already many examples, should not be taken lightly. Biotechnology can be as much a threat to natural species and genetic diversity as it is a benefit and a justification for protecting that diversity. The release of individuals with artificially composed genetic traits into a wild population disturbs the natural genetic pool of that population and affects competitive interactions, destabilizing the natural biological community—a potential problem in aquaculture. There are also concerns related to ethics, intellectual property rights, and humane treatment of animals.[20]

As already mentioned, marine ecosystems are characterized by a diversity of functions as well as a diversity of species. Some of these functions have special value to humans and are referred to by economists as *ecosystem services*. Seagrass meadows, for example, provide protective nursery areas for many estuarine and ocean fish. Healthy estuaries are nutrient rich and produce an array of food organisms that support both their own productivity and that of adjacent coastal waters; consequently, they support valuable coastal fisheries. Some coastal areas are abundant sources of larvae that are transported by currents to other areas, where they replenish depleted populations, including populations of commercially valuable shellfish. Areas of coastal upwelling provide nutrients for highly productive food webs that support other sea life and rich fisheries. Coral reefs provide physical structure, food, and protection for a great diversity of marine species. The coral itself is composed of calcium carbonate produced by animals and plants in a process that sequesters large quantities of carbon dioxide from the environment, thereby helping to moderate global warming. Marshes trap sediments and filter chemicals from the water, a function that may help protect coastal waters from some of the pollution that people carelessly allow to flow from the land—a service they should not be called on to perform.[21]

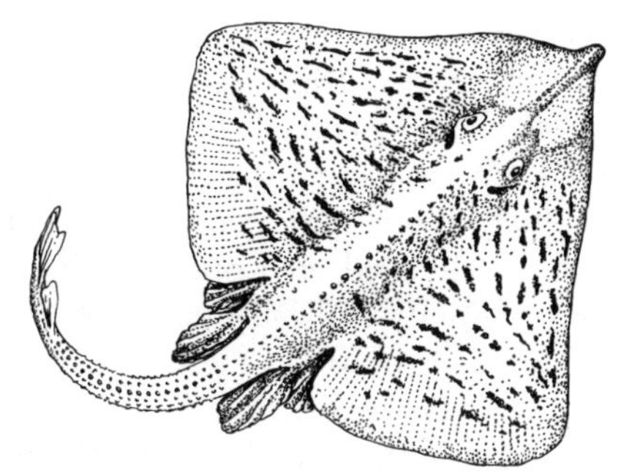

Chapter Two

The Threats to Marine Biodiversity

Biodiversity has become widely recognized as a critical conservation issue only since the 1970s with threats to marine biodiversity first receiving attention only in the 1990s. The rapid destruction of tropical rain forests, the ecosystems with the greatest known species diversity on land, awakened people to the importance and fragility of biodiversity. The high rate of species extinctions in those environments is truly alarming, but it is important to recognize the vulnerability of biodiversity in all ecosystems.

Extinctions occur continually at a low rate, known as the background rate, but in the earth's long history there have also been several periods of mass extinction during which a substantial fraction of species were lost—at least 25 percent and probably closer to 50 percent or more—over time spans of no more than 1 million years. Most famous is the late Cretaceous mass extinction, which saw the demise of the dinosaurs, among other species. Only 0.01 percent of the species that have lived on earth have survived to the present time, and it has largely been chance that has determined which ones survived and which died out. Because of the magnitude and speed with which the human species is altering the physical, chemical, and biological world, biodiversity is being destroyed at a rate unprecedented in recent geologic time—perhaps 10 to 100 times faster than ever before—placing us in the midst of one of greatest mass extinctions in the planet's history. [1]

Because there are more species now than there ever have been, the numbers lost will be greater than ever. There are predictions that within 50 to 200 years, human activity will have caused the loss of 50 percent of the species with which we have shared this planet. More than a few will be marine species. It is true that life on earth has continually been in flux as physical and chemical changes occur, but life needs time to adapt—time for migration and genetic adaptation within existing species and time for the proliferation of new genetic material and new species that may be able to survive changed environments. The swift attack on the environment by humans and resulting biological wreckage may not allow the usual compensatory evolution of new species.

Past mass extinctions have each been followed by an unusually rapid (on an evolutionary scale) invasion of species from other areas and an ensuing generation of new species, though not as rapid as the initial decline in diversity. Whether this rebound will occur after the current extinction finally slows is not clear; it is just as likely that a relatively few opportunistic species—weed and pest species—will proliferate. The current human-induced mass extinction is taking place on a much shorter time scale than the geologic time frame of past extinctions and is occurring in all environments, including the marine environment.[2]

The ocean environment is gentle and stable compared with that on land. Seasonal and yearly oscillations in the environment are muted, especially in the open ocean, far from the land's edge. These favorable conditions have supported the proliferation of a multitude of species, but many may not be easily adaptable to sharp and sudden changes in their environment on either short or long time scales.

The numerous species have sorted themselves into zones or *biomes* (ecosystem types) corresponding to sets of environmental conditions that vary with depth, latitude, and longitude. The boundaries around these zones are soft, permitting a significant cross-boundary flux of chemicals and living creatures. Some scientists have suggested that this easy accessibility to other marine ecosystems may compensate for lack of adaptability, since species can move away from deteriorating environs and seek out more suitable habitats. However, it is clear that not all species can move so readily, and many more than we realize may be threatened because they cannot adjust to rapid environmental changes imposed by human activities. Furthermore, it is now recognized that human-induced degradation of the marine realm is far more extensive and substantial than previously acknowledged by science. The major threats to marine biodiversity are summarized in the sections that follow.

Overfishing, Overhunting, and Aquaculture

The demise of marine mammals, seabirds, and sea turtles as a result of human assault is no longer a threat—it is a *fait acompli*. Many species and populations have gone extinct, and many others are impoverished or on the brink of extinction. Among the missing are the Steller's sea cow, the Atlantic gray whale, the Labrador duck, the great auk, and the Caribbean monk seal—all victims of overhunting. The hunting of whales reached its peak in the 1950s and 1960s, when the whaling industry was killing more than 50,000 whales per year, and it still continues at a reduced level. In 1982, by international agreement, an indefinite moratorium on commercial whaling was put into effect, but several nations decline to honor it—Norway, Japan, Russia, and Iceland have all violated the moratorium. Even more ominous, however, are the living conditions we have created for these animals. Depending on the species, toxic pollution accumulates in their tissues and causes reproductive abnormalities and deformities; diseases run rampant in their populations; red tides poison them in large numbers; abundances of forage fish are greatly reduced; breeding areas are disturbed; and ships ram into them. Furthermore, modern fishing technologies, which indiscriminately strip the seas of everything above a certain size, kill countless marine animals. With all we are doing to make life difficult for whales and other endangered marine animals, the purposeful large-scale hunting of those few that still have viable populations seems immoral as well as senseless.

Fisheries have also reached a level of incomprehensible destruction. Like the proverbial dog that bites the hand that feeds it, commercial fishers around the world are rushing to scoop out the final remains of what once seemed an endless supply of protein provided by the ocean. Fishery species have always been viewed as independent stocks, never as integral parts of organized ecosystems, and that undoubtedly has been their downfall, for those "stocks" are now subject to crashes worldwide. Coastal ecosystems, not surprisingly, have proven susceptible to the combined effects of overfishing, coastal pollution, and habitat alteration, and the fisheries themselves are vulnerable to other stresses.

Coastal fisheries are generally seasonal and may supply local, regional, or global markets. Depending on the extent of national government regulations, many coastal fisheries have been depleted or have been temporarily closed due to threat of depletion. Commercial fisheries farther offshore in international waters are not subject to national regulations and only recently have come under scrutiny. Whereas coastal fisheries exist to some extent in nearly every coastal country, open-ocean fisheries involve only a few nations, but they serve global markets. The operations are large, and fleets are technologically sophisticated. A factory trawler in the North Pacific, for instance, typically brings in 300,000 to 1 million pounds per haul. Many modern fishing vessels are floating factories, handling gear of huge proportions: perhaps nearly 100 miles of longlines with thousands of hooks that catch anything from the target fish to seabirds, or perhaps a trawl net large enough to hold a dozen jumbo jets, or fifty-mile-long drift nets. The large number of vessels using this technology creates a fishing pressure so great that 80–90 percent of some fish populations are killed every year. At that rate, they don't recover very easily. [3]

Although several thousand species of fish have been hunted worldwide, fewer than thirty are taken in quantities that supply global markets, and of these only five account for nearly half that market. With too many boats and too much high-tech gear, the fishing fleets of the world fished until the annual yield of these species—after increasing rapidly through the 1960s and 1970s—slowed in the 1980s and began leveling off and perhaps falling in the 1990s despite increases in the size of the fleets. The Food and Agriculture Organization of the United Nations (FAO), which keeps statistics on fisheries, claims that nearly half the world's commercial fishery species are fully exploited and another 20 percent or more are overexploited or depleted. In the United States, the National Marine Fisheries Service has stated that nearly 80 percent of monitored commercial fishery populations are overfished or "fished to their full potential," including nearly all the commercially important fisheries off the Atlantic and Gulf coasts.[4]

These disturbing figures are based on the questionable notion that for every exploited species, there is an intensive level of fishing that can be sustained if it is not exceeded (referred to as *maximum sustainable yield*). Some scientists and environmentalists believe that to be sustainable, fishing must be reduced far below the present levels. In fact, the whole idea of "overexploited," "fully exploited," and "under exploited" species raises serious issues about the way we view life in the ocean. What is an underexploited species, anyway? That term implies that fish are in the ocean for one purpose only—to be exploited by humans.[5]

Far from being a minor problem, correctable by small adjustments in effort and catch, the hunting of fish by humans for commercial, sport, and subsistence purposes has led to a global crisis in marine biodiversity. Yet as bad as the situation is for those marine species purposely exploited by humans, that is only part of the damage wrought by fisheries. The rest of the story lies in the effects fishing has on other species in the ocean. First, enormous amounts of "nontarget" fish and other animals, referred to as *bycatch*, are inadvertently caught in nonselective fishing gear and discarded. Second, uncaught species are seriously affected by the removal of substantial numbers of the "target" species. Finally, bottom habitats are ripped up and destroyed by trawling gear being dragged along the bottom in estuaries or coastal ocean areas to gather one species or another.

Of the target fish, the species most threatened are those with low reproductive rates and long life spans, such as swordfish and orange roughy. They may reach reproductive maturity only when they are many years old. Because fish are often killed before they can reproduce, fewer are hatched each year and over time, the average fish in the surviving population become progressively smaller and younger. Short-lived fish may be threatened, too, as evidenced by the collapse of the sardine fishery off California in the 1940s and the collapse of the anchovy fishery off Peru and Chile in 1972. The recent collapse of the New England groundfish and Canadian cod fisheries was the direct result of too many boats, including foreign trawlers in international waters, and insufficient regulation, both national and international. Although no commercial fish species has been reported to be extinct, some populations have (e.g., certain runs of salmon on the Pacific coast, sturgeon in the Black Sea, giant clams in the Pacific Ocean). Many others are endangered, and genetic diversity has clearly been reduced in populations of numerous species. As specific fisheries are depleted, the fishing effort shifts to new species, resulting in progressive exploitation down the food chain. Fisheries commonly focus on large predatory fish; as these are depleted, smaller species lower on the food chain are taken, as are some pelagic invertebrates, such as squid. As world fisheries move down the food chain, the populations of more and more animals in the ecosystem become impoverished and systems are dominated by smaller and smaller organisms. [6]

The bycatch may be even more devastating to pelagic (open-ocean) ecosystems than the catch. In pursuing the target species, modern commercial fishers catch and kill numerous other species in quantities that may exceed the target catch. These include marine mammals, seabirds, and sea turtles, which are often protected by law, as well as numerous other species of fish and invertebrates. Some of the unwanted catch is converted into fish meal; the other animals are discarded, dead or dying. The effects of these practices on populations and whole ecosystems are difficult to assess. Pelagic, demersal (living near but not on the ocean floor), and benthic (bottom-dwelling) species are at risk. Commercial fishers in the pelagic fisheries have caused tremendous devastation by purposely or accidentally releasing nets. Drift nets, now banned in many fishery areas, are nets miles long that are released for a period of time to catch whatever they come upon and then hauled in with their mixed catch. Lost "ghost" nets still roam the seas reaping their grim harvest and leaving pelagic communities severely impoverished. Finally, trawlers dragging the bottom for demersal and benthic fish and shellfish destroy entire sea-bottom communities in their paths. Such waste is, in effect, subsidized by the high prices fetched by some of the target fish, prices that escalate as individuals of the species become harder to find.

There are significant secondary effects of the changes in populations of captured species. The large predatory fish that are favored often play a key role in regulating species diversity in their ecosystems. For instance, overfishing of grouper on coral reefs has resulted in overgrowth of the algae that the grouper would have eaten. In another case, seals killed by commercial fishers to keep them from eating cod had in fact been eating predators of the cod; in the absence of the seals, the cod declined further. And in Delaware Bay, a decrease in the population of horseshoe crabs, which are taken for bait, has caused a decline in the populations of shorebirds that rely on the horseshoe crab's eggs. Another secondary effect of population changes in fished species may be a change in competitive interactions as new species move in and others disappear. Cascading effects may be expressed as shifts in species throughout the food web, and species and genetic diversity in general may decline. Furthermore, there may be serious implications for the genetic diversity of any of the overfished species. For example, selective removal of the largest fish in a population may favor such characteristics as smallness and early maturation so that reproduction occurs at a smaller size. A severely reduced population will also have less genetic variation simply because the opportunities for variation are reduced. Thus the biological character of many marine ecosystems has been fundamentally and perhaps permanently altered by the careless and overwhelmingly aggressive methods of modern fishing fleets.[7]

Physical destruction of the ecosystem is commonly associated with fishing by use of explosives in coral reefs and with dredging, trawling, and long-hauling (dragging nets along the bottom) to gather bottom fish or shellfish. These practices are devastating enough when used once, but when used repeatedly in the same area, they literally wipe out the benthic and demersal community, and any pelagic species that rely on that community will move out of the area. The loss of species in such an ecosystem may be enormous. Commercial fishing has truly become the great impoverisher of marine ecosystems worldwide.[8]

As fisheries decline, many countries have turned to aquaculture to supplement their production of marine fish and shellfish for the global market. In theory, aquaculture holds great potential for reducing fishing pressure on wild populations and for enhancing coastal ecosystems. In practice, however, modern marine aquaculture, a high-profit activity not aimed at providing food for hungry people, almost invariably harms the coastal environment and threatens wild populations. The problems are many and include destruction of coastal habitat, spread of fish disease, introduction of new species or new populations into the natural environment, and pollution, as well as numerous human social problems. All these threaten natural biodiversity in coastal areas where aquaculture is practiced. Problems arise because the technologies and facilities are not designed with environmental protection in mind and there is no attempt to select species that are compatible with the natural ecosystem or to cultivate balanced mixtures of species.[9]

Physical Destruction of Habitat

One of the most obvious sources of biodiversity loss is the physical destruction of significant expanses of natural habitat. Such destruction has ruined or degraded countless ecosystems on land and is the major threat to tropical rain forests, but it is taking a toll on marine environments as well. The coastal zone is the most vulnerable, for it is here that humans build on, cut through, dredge out, or bury entire marine communities. With an estimated

37 percent of the world's population living in coastal areas within 100 kilometers (62 miles) of shore, the pressures of coastal development are severe. Loss of critical coastal habitats is caused by a mélange of coastal construction and modification projects, including port and harbor development; dredging of navigable waterways; road and railroad building; filling, diking, and channelization of wetlands; building on barrier beaches and sandy sea cliffs; construction of seawalls, groins, and jetties; excessive sedimentation eroded from upland clearing and construction; coastal mining; certain fishing and aquaculture operations; construction of artificial islands for airports, ship terminals, and disposal of dredged materials; development of marinas and other waterfront tourist facilities; and reduction of water inflow from rivers due to dams and diversions. The United States had lost roughly half its original coastal wetlands by the mid-1970s. Similarly, mangrove ecosystems in tropical countries face tremendous pressures when they are overused as a source of firewood and food, are cleared and converted to brackish-water shrimp ponds, or are leveled for housing or industrial development.[10]

Trawling and dragging of the floors of estuaries, seas, and coastal waters have had profound effects on inshore and offshore bottom habitats. Repeated clearing of bottom communities in pursuit of these activities has devastated the sea-bottom ecosystem. Recovery is impossible because there is not enough time for the animals to get fully reestablished before they are cleared out again. Similarly, mining of the sea floor destroys the bottom-dwelling animal community and increases the load of sediments, possibly burying some habitats.

Reduction of the amount of freshwater flowing from rivers into estuaries has become a serious cause of habitat destruction. This is the result of inflowing rivers being diverted or dammed to hold back water for irrigation and human consumption. The result is increased salinity and saline intrusions into freshwater, reduction in delta size because of reduced sediment loads, changes in temperature, changes in nutrient dynamics, and increased concentrations of pollutants in estuaries. The distribution of species changes when an estuarine environment is confined to smaller areas, with new salinity and temperature gradients and more pollution. Dramatic examples of systems stressed by reduced water inflow include the Black Sea, the northern Gulf of California where the Colorado River used to flow in, the Nile delta, and San Francisco Bay, to name only a few. When these changes come about, the natural species diversity in the estuarine habitat is reduced by loss of submerged habitat, exceeding of salinity tolerances, exposure to toxic pollutants, and/or progressive loss of interdependent species.[11]

A somewhat different kind of physical threat comes from marine debris, including plastics, fishnets and lines, and many other forms of trash and garbage. Seagoing vessels and seashore activities are sources of this enormous problem, which can be found even in the most pristine environments, such as the coast of Alaska. The lives of fish, sea mammals, seabirds, and sea turtles are threatened by debris that may entangle and drown them or, if ingested, poison or choke them. It is not known to what extent minute particles threaten smaller forms of sea life, but laboratory experiments have shown that small particles of plastic, such as Styrofoam beads, can clog filtering mechanisms, damage digestive tracts, or at the very least deprive an animal of nutrition by interfering with the capture of real food particles.[12]

Chemical Pollution

Water pollution is more subtle in onset than is overt destruction of habitat, but it is equally devastating to the biodiversity of affected ecosystems. Furthermore, because of the fluid and dynamic nature of the ocean, chemicals entering marine waters can be broadcast far beyond their sources of emission. In addition to physical dispersal, there is biological dispersal through food webs. A significant proportion of the flux of chemicals through an ecosystem and through the global environment is determined by natural progression along food chains. This is true of chemicals that are important to the growth of organisms as well as chemicals that hinder growth or are otherwise harmful.

About 20 percent of all ocean pollution is due to activities at sea—vessel traffic, waste disposal, oil and gas exploration, and mining on the deep-sea bed. Most of the remaining 80 percent comes from activities on land. Land-based pollution is discharged or runs off into the sea or into rivers flowing to the sea or is emitted into the atmosphere and carried on wind currents to the sea surface. Not only are most ocean waters along developed coasts polluted to some extent, but contamination has also reached the middle of the ocean, the deep-sea floor, and coastal waters far from any sources of contamination.[13]

It is the very nature of water that makes pollution such a threat in marine environments. The chemical outfall of human civilization is soluble and reactive in water. Consequently, organisms that live and feed in water are highly susceptible to the effects of foreign substances or an overabundance of familiar substances. In water, the chemistry of pollution reacts readily with the chemistry of life, and the two don't mix well. Chemical pollutants fall into two major categories—excess nutrients and toxic substances—which differ considerably in their fates in the marine environment and their effects on organisms and ecosystems.

Before further discussing these two types of pollution, terminology should be clarified. *Pollution* is a functional word implying negative biological effects. *Pollutants* are substances known to cause pollution, and *contaminants* are any chemicals in the water that do not occur there naturally, whether or not they result in pollution. The term *toxic* does not mean that the substance always causes death; other harmful and chronic effects on living organisms are also common. Toxic substances may be natural or synthetic.

Nutrient Pollution

There are many minerals that are essential to the growth and health of living organisms—nitrogen, phosphorus, silicon, and iron, to mention only a few. The ocean is a solution of many salts that contain most of these nutrients, but a few, such as the four just mentioned, are sometimes in short supply, limiting the growth of algae, the plants of the sea. Whenever significant supplies of these limiting nutrients enter ocean waters, they cause a burst, or bloom, in the growth of algae. Most significant in this process are the microscopic plants known as microalgae or phytoplankton that live in the sunlit waters near the sea surface.

The influx of nutrients may be natural, caused by storms stirring up nutrient-rich waters from deeper in the sea or flooded rivers carrying out eroded sediments rich in decayed plant material and therefore rich in nutrients; or it may be augmented by human activities, such as runoff from agricultural lands laden with fertilizers and animal wastes, discharge from sewage treatment plants or seepage from septic systems, or airborne emissions from the burning of fossil fuels that are carried by winds and deposited on the sea surface. Air pollution contains nitrogen-rich compounds that fall on estuaries and coastal waters, increasing the concentration of nitrogen available to the food chain. These same compounds are in acid rain, which acidifies freshwater lakes and streams. The salt solution of the ocean is so highly buffered that acid rain does not cause marine acidification but the added nitrogen leads to nutrient pollution. As much as 25 to 40 percent of the nitrogen in coastal waters may come from the air. In Chesapeake Bay in the United States, one of the first places where this was recognized, at least a quarter of the nitrogen entering the bay originates in air pollution from automobiles and coal-burning power plants as far away as Ohio and perhaps beyond.

There is an important distinction between nitrogen gas, which is a major component of the earth's atmosphere, and the airborne nitrogen compounds that ionize when they dissolve in water and are in a form that plants can use. The gaseous nitrogen molecule is not readily usable by most living organisms, although there are certain "nitrogen-fixing" bacteria that can capture it; just as there are "denitrifying" bacteria that release gaseous molecular nitrogen. Dissolved nitrogen-rich pollutants, however, are used directly by the major primary producers in the ocean—the microalgae, blue-green algae, and prochlorophytes (discussed further in Chapter 3).

Nitrogen is perhaps the most significant nutrient pollutant arising from human activities—because we are releasing so much of it into the environment and because it is often the limiting nutrient in coastal waters. Initially, the addition of nitrogen will merely boost the growth of whatever organisms are present, but eventually those species able to use it most rapidly will bloom and push out other species, thus reducing species diversity, an effect that cascades through the food chain. Fertilizer production and application and other human activities have more than doubled the global rate of input of biologically usable nitrogen into the environment. Much of this makes its way to the ocean, where it enriches coastal waters to the detriment of the ecosystem.

A bloom of algae resulting from an influx of nitrogen or other nutrients will, up to a point, fuel a healthy increase in productivity of the whole ecosystem without sacrificing biodiversity over the longer term. However, as human sources of nutrient pollution increase, nutrients may accumulate in very high concentrations in estuaries and nearshore waters, and when this happens the bloom may take on a different nature, severely affecting the whole ecosystem. The growth of algae outstrips the ability of grazing marine animals to consume it, and as the bloom gets denser and shades itself from sunlight, algae die and sink to the bottom. The bacteria that decompose the dead algae use oxygen dissolved in the water, which leads to low oxygen levels (hypoxia) or an absence of oxygen (anoxia) in the bottom waters. This obviously is not good for animals living on or near the bottom; they will leave if they are mobile or suffocate if they are sedentary. At every level of the food chain, the diversity of species decreases and the ecosystem deteriorates. This process is known as *eutrophication*.

The damage is reversible if the nutrients return to healthful levels, water circulation reoxygenates the water, and plant and animal species move back in. However, some estuaries and coastal waters are so constantly polluted with nutrients that they never recover. Estuaries and other waters with restricted circulation are the most susceptible. However, one of the worst examples of eutrophication is located in shallow open waters of the Gulf of Mexico. The 18,130-square-kilometer (7,000-square mile) "dead zone" off the mouth of the Mississippi River, which carries an enormous load of nutrients from the agricultural lands of the central United States, is an area of eutrophication where the waters becomes hypoxic or anoxic in the summer. This area has progressively grown in both size and duration of the hypoxic conditions. The area's name says it all—living creatures die and biodiversity is severely reduced there. The New York Bight, which receives the outfall of several northeastern cities, is also affected by eutrophication and has periods when waters are hypoxic, so its ecosystem also has suffered loss of biodiversity over the years. In areas such as this, however, where there are so many sources of different kinds of pollution, ecosystem changes cannot be attributed to a single cause.

Nutrient pollution probably plays a significant role in the promotion of harmful algal blooms. These are blooms of algae that produce toxic metabolites, are not a good food source for many grazers, or in some other way cause damage to the food chain. In the most general sense, eutrophication is harmful and therefore the bloom that leads to oxygen depletion can be termed a harmful algal bloom. More commonly, however, the term refers to a bloom in which a single species of microalga takes over and becomes directly harmful to its predators and/or other animals in the ecosystem. Sometimes the word *harmful* is laden with human values; sometimes a particular bloom is more harmful to human economies than to the long-term fate of the ecosystem. Nutrient pollution generally leads to a reduction in phytoplankton species, but in extreme cases the excess nutrients combine with other environmental properties to create the perfect conditions for one opportunistic species to bloom at the expense of other species. Such blooms are often called red tides, brown tides, or green tides because they grow so densely that the water may take on the color of the species that is blooming. These blooms are discussed further in Chapter 3.

Nutrient pollution is the only type of pollution for which "dilution" is indeed the "solution." Whether excess nutrients are helpful or harmful depends on their concentration in the water. For this reason, estuaries and closed and semiclosed seas, where circulation is restricted, are the most prone to nutrient pollution. In areas where nutrient outflow is sufficiently great, such as the Mississippi River area of the Gulf of Mexico and the New York Bight area of the Atlantic, more open coastal waters may also suffer. Generally speaking, however, once nutrients reach the open sea, they are dispersed and diluted below problematic concentrations.

This principle has often been used to justify the disposal of nutrient-rich sewage into the ocean and the open water placement of sewage outfall pipes or sites for disposal of sewage sludge from ships or barges. However, in many cases, discharge or dumping in coastal waters over long periods has resulted in eutrophication, increased cloudiness of the water, contamination by pathogens, movement of pollution toward shore, and other unwanted effects. The dispersal of dissolved nutrients is not the only issue; problems are exacerbated by the particles and toxic substances that are present in sewage sludge. In many coastal locations, the only way to achieve a more dilute solution of nutrients is, for example, to restrict the use of fertilizers, reduce airborne emissions, institute more advanced treatment of sewage (only tertiary treatment removes the dissolved nutrients), convert sewage sludge from treatment plants into toxin-free and pathogen-free fertilizer and return it to the land, or avoid using waterstream waste collection altogether by composting toilets instead of septic tanks or sewage systems.[14]

Several international institutions (international lending banks, various treaties, etc.) have identified nutrient pollution as one of the most serious marine environmental problems in the less developed countries of the world, where sewage treatment is minimal or nonexistent and burgeoning coastal populations cause heavy loads of human wastes to be washed into waterways and the sea. Nutrient pollution is also now suspected to be a major factor in the decline of coastal fisheries. Nevertheless, the importance of toxic pollution, which often occurs via the same wastestreams, should not be underestimated.[15]

Toxic Pollution

There are some chemicals that, unlike nutrients, have negative effects on life in almost any concentration and are not beneficial under any circumstance. Whether they can be tolerated by the living world at "low" concentrations varies with the substance and with the scientist studying it. In other words, there is disagreement in the scientific community as to whether low levels of toxic substances are harmless to aquatic communities and, if so, just what the limits of tolerance are.

An enormous quantity and variety of synthetic chemicals are released into the environment every year and become potential causes of environmental degradation. Some naturally occurring chemicals can also have toxic effects on living organisms, especially when they become unnaturally concentrated in the environment as a result of human activities. The range of effects of toxic chemicals on individual organisms includes deformities, diseases, lesions, sex changes, reproductive failure, interference with communication, and behavioral abnormalities, as well as death. Such effects lead to loss of genetic and species diversity—that is, biological impoverishment—in affected ecosystems.

Naturally occurring substances that may pollute marine ecosystems and harm living organisms include oil, certain toxic metals, and radioactive atoms, or radionuclides. Oil is pumped from the ocean floor, transported in tankers from land areas where it is produced, and stored in tanks of various sorts on land, often near coasts. All these are potential sources of marine oil pollution. Toxic metals may become concentrated in the marine environment as a result of runoff from abandoned or ongoing mining operations, discharge from industries that use these metals, discharge from sewage treatment plants, and urban runoff of rainwater containing dissolved chemicals from the surfaces of buildings, roads, and vehicles. Natural radionuclides are leached from mining operations.

The physical and chemical characteristics of persistent synthetic organic chemicals and toxic metals cause them to concentrate in two places in the ocean. They are both interface areas: the *sea surface microlayer,* where the sea meets the atmosphere, and bottom sediments, where the sea meets submerged land. Furthermore, ocean and atmospheric circulation patterns cause these persistent toxic chemicals to accumulate in exceptionally high concentrations in Arctic regions. Organisms living in sediments, surface waters, and the Arctic are vulnerable to high doses of these toxins, which also become concentrated in living tissue.

The sea surface microlayer is made up of organic molecules that concentrate at the water's surface, forming a film about 0.05 millimeter, or about two one-thousandths of an inch, thick that is held in place by the powerful surface tension at the water's uppermost boundary. Molecules of nutrient-rich organic chemicals, toxic metals, and synthetic organic pollutants concentrate there. Originating from waste disposal, agriculture, fuel and coal combustion, oil drilling, and other sources, organic contaminants in polluted waters beneath or in air above become concentrated in this film. Metal ions, some of them toxic, may cling to natural or synthetic organic molecules. This dense soup of both nutritious and poisonous substances may contain concentrations of toxins anywhere from two to thousands of times greater than those in the water beneath or the air above. Microscopic life also concentrates in the microlayer. It is an important area for the early development of many fish and other species with planktonic life stages and for microscopic marine species that live at the sea surface, sometimes clinging to the surface film. Effects of contaminants on eggs and larvae in the sea surface include mortality, malformation, and chromosome abnormalities all of which have been observed in such fish as Atlantic mackerel and flounder. Consequently, populations and diversity may be affected not only in the microlayer itself but also in far-flung and deep ecosystems where the adults of these eggs and larvae live.[16]

Contamination present in bottom sediments may have been there for a long time, the remnants of bygone industries, spills, dumping activities, and shipping, or it may be recent and ongoing. Because of the physical and chemical nature of some sediments, many toxic chemicals will cling to them and accumulate in concentrations very much higher than in the overlying waters. Although this may serve to isolate the toxic materials from the environment for a time, the sediments' capacity to hold on to the contaminants is finite, and eventually they slowly seep into the water as it moves over the sea bottom. The contaminants may also be ingested or absorbed by animals that feed in the mud or otherwise come in contact with it. These chemicals may be stored, to various degrees, in body tissue, where they concentrate to higher and higher levels over time in a process called *bioaccumulation*. They may also become more concentrated as they are passed along the food chain—a process called *biomagnification*.

The fate of a chemical after being ingested by an organism is determined by the balance among uptake, storage, metabolism, and elimination. Because this can vary widely with the species of organism and with the chemical in question, the amount of bioaccumulation and persistence may differ within as well as among taxonomic groups and trophic levels (positions in the food chain). Thus, biomagnification may or may not be significant, but when it is, the top-level predators, such as swordfish, albatross, and seals, contain the greatest accumulations of persistent pollutants.

Oil

Oil naturally seeps slowly from "leaks" in the ocean floor above natural oil deposits. At these sites of seepage are microbes and animals that not only tolerate the oil but even consume and metabolize it. There are not many such species, but they are an example of how life can evolve to adapt to unusual conditions.

Oil causes trouble when it is discharged in large quantities into ecosystems containing life-forms that are not adapted to its chemistry. It is lethal to many species and causes chronic health and reproductive problems in others. Large discharges of oil into marine environments are the result of spills, routine flushing of the tanks in oil tankers, or runoff from oil spills on land. Although oil spills do not account for the greatest total input of oil into the sea, they cause the most dramatic effects because the influx is so sudden and so concentrated.

Oil spills from tankers or from blowouts associated with underwater oil drilling are immediately lethal to ocean wildlife and can in short order reduce the populations of many of the species in the affected area. These disasters receive a great deal of publicity, and the dead and dying animals attest to the toxicity of the substance, but many of the deaths and sublethal effects are unseen and long lasting. Great efforts are made to de-oil birds and otters, but with minimal success—most of them probably do not survive after they are released. The effect an oil spill has on the public is evidenced by the long-lasting familiarity of the names of tankers or oil rigs long after a disaster—names such as *Exxon Valdez, Torrey Canyon, Amoco Cadiz,* and *Ixtoc* (three tankers and a Gulf of Mexico oil well associated with oil spills.[17]

Yet the effects on animals and microbes that live in and under the water are not visible to television cameras and are very difficult to study over the long term, even with as much money and effort as have gone into follow-up studies of the *Exxon Valdez* spill in Alaska in 1989. Whenever an ocean oil spill occurs, it is considered a threat only if it ends up on the coast because that is where people actually witness the mess and devastation. If it stays far offshore, it can be swept under the waves, so to speak, because no damage is seen and no attempt is made to measure it. Furthermore, after any oil spill, the oil gradually disappears from view. First, a large fraction of it evaporates as toxic fumes (which threaten sea mammals and turtles); then some of the remainder breaks up, forms tar balls, is washed away, and is slowly decomposed by microbes that can use it as a food source.

However, while this is happening, some of the oil is consumed by creatures that don't survive its toxic effects, and some kills eggs, larvae, and adult animals that cannot tolerate its chemical and physical effects on gills, cell membranes, and other structures. Some of the oil settles into the sediments on the coast and sea bottom, where it can be a source of toxic pollution for years to come. In the worst cases, it can form an underwater pavement. The total extent of damage from a given oil spill is never known, and the long-term effects are rarely documented. Clearly, spills add to the environmental stresses that together lead to biological impoverishment, and while an ecosystem is recovering from an oil spill, it is more vulnerable to other threats.

Although oil spills are dramatic, chronic operational oil discharge is responsible for more widespread damage. Millions of marine birds have been killed worldwide during the twentieth century after being coated by oil that was released into the ocean. These releases may not be large, but they are common.

Toxic Metals and Radionuclides

Several toxic metals occur naturally in seawater. Although they usually exist at subtoxic concentrations, they are often augmented by human activities to the point of being problematic. For example, metals associated with mining, industry, and waste incineration include mercury, cadmium, lead, zinc, copper, chromium, tin, and manganese; arsenic and selenium from other sources may sometimes be of concern. Several naturally occurring radionuclides—radioactive isotopes of potassium, rubidium, thorium, and uranium—may be associated with runoff from the mining of ores. These may accumulate in high concentrations near the source of contamination but are of less concern over broader areas than are artificial radionuclides, discussed later in this section. Radioactive forms of carbon and hydrogen originate in the atmosphere through natural and human-induced processes and may be dissolved into ocean waters and incorporated into the food chain.[18]

Metals are also deposited on the ocean surface from the atmosphere. Natural sources of airborne metals include wind-borne soil particles, volcanic eruptions, sea-salt spray, forest fires, and biological aerosols; but, for the most part, the amount of toxic metals from natural sources is small compared with the amount from human sources. Toxic metals are introduced into the atmosphere by factory emissions, automobile exhaust, burning of coal and wood, waste incineration, volatilization from dumps, and spray paint application. Mercury is a special case because it is so volatile that the atmospheric route to the ocean is the major concern. For this reason, mercury and many other metals are not strictly a coastal threat, as some would suggest they are.[19]

Generally speaking, toxic metals have threshold concentrations below which they do not cause damage to exposed biota. However, bioaccumulation and biomagnification may render the concentrations within living tissue much higher than in the surrounding water. A portion of the metal contamination in sediments is considered unavailable to the biota because it is so tightly bound to the sediments. However, many factors make the bioavailability difficult to accurately assess. Some animals eat a lot of sediment as they feed on the tiny organisms that live in it; the movement of water along the bottom increases the amount of materials that can be dissolved from the sediment; and tissues in direct contact with sediment may actively take up substances adhering to it.

There is not much clear documentation of the effects of metals and radioisotopes on organisms, though in high enough doses many of them can be lethal, and in lower doses over long periods reproductive and behavioral effects have been suspected. Loss of biodiversity in association with long-term metal pollution has also been documented. The effects will certainly vary from one species to another and even from one individual to another, depending on the metal involved and how the organism eliminates it, stores it, or metabolically converts it to another form. For instance, the biological methylation of mercury renders it more toxic in some animals.[20]

Synthetic Organic Chemicals

The final major class of toxic pollutants, synthetic organic chemicals (manmade chemicals containing carbon) is composed of a huge variety of substances that are produced by industry or are by-products of industrial processes. It is estimated that more than 80,000 chemicals are manufactured and about 3,000 of these account for 90 percent of the total production. It is also estimated that 200–1,000 new chemicals enter the market each year. These include pesticides, cleaning agents, pharmaceuticals, and numerous other chemicals for industrial uses. One large group of synthetic chemicals is hydrocarbons derived from petroleum; another large and highly toxic group is organohalogens, primarily organochlorines, based on chlorine chemistry.[21]

Many toxic cleaning compounds and other products of the petrochemical industry pollute rivers and coastal waters. In addition, an array of polyaromatic hydrocarbons (PAHs), by-products of the industry and of the combustion of fossil fuels, are common in coastal marine sediments. These are not truly synthetic compounds, but human industry is the main source of them, so they are included here. PAHs are natural products of forest fires, but that input has been overtaken by the input from industry and power generation. They are of concern as mutagens and carcinogens.

The greatest attention has been paid to persistent organic pollutants (known in international regulatory circles as POPs). Most are organochlorines, which include the infamous pesticides DDT, toxaphene, aldrin, dieldrin, and chlordane; PCBs (polychlorinated biphenyls) used in industries;

and dioxins and furans, by-products of the industrial use of chlorine chemistry and therefore by-products also of municipal and medical incineration. These chemicals are among the most persistent; they last indefinitely and move long distances in the environment. POPs have been found in exceptionally large quantities in the Arctic environment, where they have accumulated over decades, traveling on air currents and ocean currents from their points of release in countries at lower latitudes. All these waterborne and airborne contaminants are found in significant quantities in many marine ecosystems, sometimes far from where they first entered the environment.

Persistent organochlorines accumulate in living organisms and have been clearly associated with the full range of toxic effects on marine animals: deformities; diseases, including cancer; reproductive and behavioral abnormalities; as well as death. Some of these compounds can act in very small concentrations to interfere with hormonal pathways in exposed animals, including human beings. In many cases, they mimic estrogens. Depressed fertility and sex reversals have been found in fish and other aquatic animals.[22]

Because of the persistence, bioaccumulation, and health effects of organochlorines in marine and other environments, many pesticides now being manufactured are of another type, which includes organophosphates and carbamates. These and other degradable synthetic chemicals are not very well monitored because the assumption is made that if they do not persist, their biological effects are of little concern. Nevertheless, they may be highly toxic for the short periods of time when they are active—some of them very much more toxic than the organochlorines they are replacing. A dramatic example of this occurred in a bay on the coast of Texas, where a large dolphin kill was attributed to aldicarb (a carbamate pesticide) runoff. Furthermore, though these chemicals break down quickly, little is known of the effects of some of the breakdown products.

Given the known effects of organic contaminants, their mode of action, the small concentrations at which they can have an effect, and their incomprehensible diversity, they pose a clear threat not only to individual organisms but also to species, ecosystems, and biodiversity in general. The synergistic effects of the soup of unnatural chemicals that enters the marine environment daily can only be guessed, but they are clearly contributing to the biological impoverishment of marine ecosystems.

Artificial Radionuclides

Nuclear weapons testing results in the emission of a variety of artificial radionuclides into the environment, and the fallout from these tests represents the largest source of artificial radionuclides in the ocean. However, these emissions are being restrained by international demands. Other sources, both historical and current, include the dumping of radioactive waste from weapons production and nuclear power plants and the reprocessing of spent nuclear fuel elements from reactors. And, of course, there continues to be the threat of accidents at nuclear reactors, such as the major release of radionuclides into the atmosphere from the accident at the nuclear power plant in Chernobyl, Russia. During the mid- to late 1990s, information came to light on the huge extent of disposal of nuclear wastes in the rivers and seas of Russia. These materials have entered the food chain, and the people of Arctic Russia, whose diets rely heavily on seafood, have experienced serious health effects. When these materials enter the marine environment, some quickly adhere to sediments and do not move long distances from their source, but others may remain in the water for much longer and travel much farther. For example, radionuclides from a reprocessing plant in the Unites Kingdom have been traced to ocean waters around Greenland. The effects of these contaminants on marine animals are not well known, and their effects on biodiversity have not been documented.

Introductions of Species

A variety of human activities are promoting the transfer of species from one location to another, and other activities favor the explosive growth of a few species at the expense of others in affected ecosystems. These two circumstances sometimes coincide, with disastrous results. One or more introduced species may flourish in a new habitat, growing so vigorously as to drive out or diminish some of the native species. Although it is possible for this to happen in healthy ecosystems, the chances of a new species becoming wildly successful are multiplied many times when the receiving ecosystem is already damaged and/or missing some of its normal complement of native species. The geologic record reveals periods when mass invasions of species became established in new ecosystems that had previously suffered a serious decline in native species due to some environmental change or disaster. This, of course, happened on a geologic time scale.[23]

Scientists warn that the rapid transfer of species in the marine environment has reached catastrophic proportions, propelled by the routine discharge of ballast water by ships in ports around the world, many of which are located in estuaries. After a ship's cargo is unloaded, seawater may be pumped into its ballast tanks to compensate for the loss of mass. The ship then sails to another port, possibly far across the ocean, where it releases the ballast water before taking on new cargo. Thus, every day, thousands of species are introduced into new environments, where they may die, become minor constituents of the community, or thrive and overwhelm the native biota. Geographic separation dictates that the native flora and fauna of distant estuaries will not be the same; but when two estuaries are similar in temperature, salinity, and water dynamics, the receiving environment may favor the introduced species. [24]

There are numerous examples of species introductions that have been catastrophic. The introduction of the comb jelly into the Black Sea, presumably from Chesapeake Bay, developed into a massive invasion. The zebra mussel flourished in the Great Lakes and spread throughout freshwater systems in the United States after it was introduced by ballast water from an estuary in Europe. San Francisco Bay has been particularly susceptible to visitors that become permanent residents, including the now prevalent Asian clam and the aggressively successful green crab. It is important to note that all these areas were stressed by pollution, overfishing, and/or water diversions when the offending non-native species invaded. Introduced species themselves may add to the stresses on an ecosystem and compound the vulnerability of an area to invasion. San Francisco Bay is also characterized by a very young natural community—it is geologically young, and in 1862 it was subjected to a flood that apparently wiped out several species that could not survive the freshwater conditions. It may, therefore, not have a full complement of native species performing all possible roles in the ecosystem. Thus, open niches may favor introduced species, allowing them to flourish.[25]

Species have always invaded ecosystems—that is how new islands become populated with life, and it is how changing environments evolve new complements of species better adapted to new conditions. Succession of species is a natural process and is important for the perpetuation of life and the ability of ecosystems to adapt to change and maintain complex biological communities. However, the artificially high numbers and varieties of species being introduced are initiating ecosystem changes that are destructive and may severely threaten biodiversity, as has happened in many locations. One simple example is the effect of domestic cats on coastal bird populations.

Though not nearly so massive as ballast water introductions, another source of introduced species is mariculture (marine aquaculture). Sometimes species are intentionally released into coastal estuaries to grow and be harvested; for example, oysters from the East Coast of the United States and from Japan have been grown in natural areas along the West Coast. Sometimes aquaculture animals are grown in pens and some escape, as has happened with the Atlantic salmon in Washington State and British Columbia. Sometimes other species are introduced along with the cultured species—these may be marine plants used to pack eggs or young animals when they are shipped to the culture facility, or they may be species incidentally growing on the shells of cultured shellfish, or they may be parasites or diseases of the cultured species.

There is no question that the tremendous quantity of species being moved around the globe by human activities is a reason for concern, but the response to the problem has been just as troublesome. Lawmakers and environmental managers have waged war on all non-native species except those that are purposely introduced for economic gain. Although little is done to prevent the introductions, an arsenal of deadly weapons is unleashed against them once they become established. Unfortunately, this arsenal, which consists primarily of numerous pesticides and may include physical destruction of the ecosystem, is not species specific and destroys native species along with the invaders. Government agencies wage this war in a strange and often emotional xenophobic reaction to foreign species. Yet economically valuable species are, with the blessing of government agencies, imported, planted, grown, and harvested, often at the expense of native species and with the intent that they become established members of the natural community. Furthermore, although experimental methods are being tested to deal with the introduction of species by release of ballast water, progress is very slow, and for economic reasons there is little attempt to restrict port activities.

There is one opportunistic species that has successfully invaded nearly every coastal marine ecosystem, invariably causing a reduction of natural biodiversity. This species is a top-level predator with effects that cascade throughout the food web. It also produces potent toxins that have a variety of harmful effects on a broad spectrum of life; and it physically disrupts the environment. We know more about this invader than we know about any other marine species, yet we have not been able to stop its progression. In fact, it is a terrestrial species that has proven incredibly inventive in its ability to exploit the marine environment—*Homo sapiens*.

Global Atmospheric Change

Human-induced global climate change also threatens marine biodiversity. There is little doubt that global warming is taking place as a result of the accumulation of carbon dioxide, methane, and other greenhouse gases. The breakdown of stratospheric ozone due to continued accumulation in the upper atmosphere of certain widely used volatile synthetic chemicals, especially the chlorofluorocarbons (CFCs) used as refrigerants, cleaning solvents, and aerosol propellants, also threatens marine organisms living near the surface of the sea.

Atmospheric warming is expected to be greatest at higher latitudes in the Northern Hemisphere, but effects on the marine ecosystem are expected, and may already be occurring, at lower latitudes as well. A temperature increase of only a degree or two can have a dramatic effect on biological communities. There will also be other effects, such as sea-level rise due to thermal expansion of the oceans and melting of ice from the Arctic tundra and ice cap. Among the possible effects are significant loss of coral reefs, salt marshes, and mangrove swamps that are unable to keep up with a rapid rise in sea level; loss of species whose temperature tolerance range is exceeded, including many coral species already living at the top of their range; and effects from the tundra runoff, which would probably include reductions in salinity and increases in nutrients and suspended sediments. A potential physical effect that could have tremendous biological ramifications is a change in the dominant pattern of circulation in the world's ocean. Currently, that pattern is driven by the sinking of cold, salty waters in the Arctic Ocean near Greenland, which drives a circulation that ultimately lifts nutrient-rich bottom waters to the surface in various areas of upwelling. Models suggest that a slight warming of the Arctic water could cause a cessation in this sinking, thus stopping the "conveyer belt" so important to the cycling of carbon, nitrogen, and other elements that are essential to life. Sea-level rise will result in saltwater intrusion that wreaks havoc with freshwater ecosystems, including rivers, freshwater marshes, and coastal lowland farm acreage.

There is increasing evidence that warming of the ocean water will result in loss of biodiversity, at least in some regions. Analysis of long-term data sets monitoring physical and biological properties of the California Current between 1950 and 1993 revealed a distinct reduction in the production of plankton correlated with a warming period leading up to the 1980s. It is suspected that the decline was caused by the nutrient-poor conditions associated with the warmer waters.[26]

With respect to ozone depletion in the upper atmosphere, there have already been measured effects associated with the south polar hole in the ozone layer. Fish eggs, which often float at the ocean's surface, are very sensitive to ultraviolet (UV) radiation; the portion of the UV spectrum considered most damaging to organisms is known as UVB, which increases its penetration of the atmosphere as ozone is depleted. During the spring, when the ozone hole is large, many Antarctic animals reproduce, so numerous eggs, embryos, and larvae are in the plankton and thus are exposed to increased UVB radiation. Damage to these animals during this sensitive phase will probably have cascading effects on larger animals through reductions in the food chain. Indeed, declines in krill populations have been measured, and although the cause has not been isolated, the ozone hole and resulting increase in UVB radiation are suspected. Phytoplankton, which also live in ocean surface waters, have a demonstrated sensitivity to UVB; declines in production of as much as 12 percent have been measured in Antarctic waters in conjunction with increased UVB radiation. The Arctic ecosystem is vulnerable in the same ways as the north polar ozone hole grows.[27]

Marine biodiversity is threatened just as terrestrial biodiversity is threatened. Human excesses—the overwhelming growth of populations and the unchecked wastefulness of "advanced" societies—are causing environmental disturbances that cascade from land to sea at a perilous rate. Life in the sea is vulnerable to the environmental consequences of the way we live.

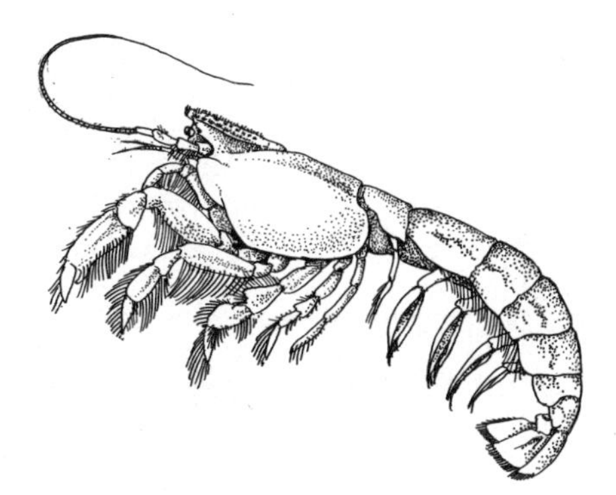

Chapter Three

How Science Describes Marine Biodiversity

Policy and regulatory decisions regarding the conservation of biodiversity are based on available scientific information. It is both the wonder and the curse of science that the more information that is amassed, the more uncertainties that become apparent and new questions that arise. It can be frustrating to citizens and policy makers who are trying to understand scientists, but sometimes it is the uncertainties that must guide policy. Careful scientific assessments of ecosystems and their species diversity by scientists familiar with ecological principles and uncertainties can lead to informed recommendations for the most prudent policies and management decisions.

Since we do not see, interact with, or protect biodiversity on either a global spatial scale or an evolutionary time scale, it is more useful to characterize biodiversity on smaller, manageable scales. Therefore, although it is important to be aware of the larger context, it is especially important that we be familiar with the processes that regulate biodiversity within finite ecosystems or subsets of ecosystems. This chapter therefore reviews some scientific terminology, theories, and concepts relevant to the conservation of biodiversity in marine ecosystems.

Terminology Related to Species Diversity in Ecosystems

All types of biodiversity discussed in Chapter 1 are important to comprehensive conservation policies, but species diversity is most commonly the focus of scientific research and management decisions involving biodiversity. Scientists have special terminology to distinguish different characteristics and quantifiable properties related to species diversity of a community, so that they know exactly what has been measured—biodiversity is too general a term. Although it may be somewhat tedious to deal with a new vocabulary, some of the terminology is actually quite helpful in understanding how ecosystems work and what determines the diversity of species they support.[1]

To describe the complexity of a biological community, the following terms apply:

- *Species richness* refers to the number of species in a system.
- *Species evenness* is a measurement of how evenly species are distributed and how evenly the populations of each species are distributed.
- *Interspecific interactions* are direct interactions between species, such as competition and predation.
- A *diversity index* is a mathematical expression that can be used to compare diversity in different biological communities or in the same community at different times, when it is impossible to count every single species. It quantifies the relationship between the richness and evenness of species.

The interrelationships and the distribution of functions—or division of labor—among species are important factors determining the number of species in an ecosystem, and several terms pertain to that:

- A *niche* is the role or set of functions a species performs in its ecosystem; different species may have overlapping but not identical niches.
- The terms *food webs* and *food chains* describe the relationships among species on the basis of who eats whom. Within this context, species may be described as being at one *trophic level* or another—for example, primary producers (mainly plants), grazers, first-, second-, and third-level (or higher) predators, and decomposers.
- A *generalist species* assumes many functions and therefore occupies a large niche in the ecosystem.
- A *specialist species* has a relatively narrow set of functions and therefore occupies a small niche in the ecosystem.
- From the perspective of some ecologists, it appears that there is *redundancy* in ecosystems, which means that more than one species performs each function. Usually, this would be the result of species having overlapping niches. It is hypothesized that redundancy is necessary for the stability and long-term survival of marine ecosystems.
- *Reproductive strategies* are important in determining the dispersal of species. Species may be long- or short-lived, may produce numerous or

few young or larvae, and may disperse larvae over long or short distances. All these factors affect the range over which a species is distributed and the genetic composition of its separate populations.

Evaluation of the stability of ecosystems is also of interest to ecologists, as species diversity and stability seem to be interdependent. There are several ways in which an ecologist can talk about the stability of an ecosystem, each with its own terminology:

- *Stability* is a community's constancy, or persistence over time. A system is deemed stable if, following a disturbance, its functional variables and much of the species diversity in the system return to the initial equilibrium.
- *Resilience* refers to the speed with which functional variables and species diversity return to equilibrium following a disturbance; and this can vary depending on the strength, magnitude, and time of the disturbance.
- *Resistance* refers to the degree to which the ecosystem changes following a disturbance.
- *Variability* is the natural change in population densities over time.
- *Biological integrity* refers to a system's completeness, including all elements (species and functions) necessary for the system to carry out all its natural processes at appropriate rates.

The relationship between stability and species diversity of an ecosystem is complex. Generally, a greater species diversity imparts greater stability to the ecosystem as a whole. However, individual species may not be protected by being in a high-diversity ecosystem. Instead, it seems that under stress, a high-diversity system may lose more species or populations may fluctuate more, even though the ecosystem as a whole is more stable. High-diversity ecosystems contain more redundant species, which have similar or overlapping functions. This may help explain their increased stability; a loss or decline of one species is compensated for by the presence of other species able to take over its functions. Redundancy likely helps maintain integrity as well as stability.[2]

Theories Related to Species Diversity in Ecosystems

Several factors, both physical and biological, that influence the species diversity of an ecosystem have been described by terrestrial and marine ecologists. The species diversity in any particular ecosystem will reflect the interaction of factors that favor greater diversity with factors that reduce diversity. There are opposing theories on whether there is a maximum biodiversity characteristic of a particular ecosystem or whether an individual ecosystem's diversity is some subset of diversity on a regional scale and in theory can increase up to that total regional diversity. Whichever theory is correct, the fact is that similar types of ecosystems in different places around the globe are characterized by different levels of species diversity.[3]

Physical and Chemical Factors Affecting Species Diversity

- Some environments have been stable over longer periods of time than others, and therefore the evolution and proliferation of species in ecosystems with stable environments has simply gone on longer.
- The larger an ecosystem's area and/or volume, the greater the species diversity it can potentially support.
- Ecosystems characterized by greater physical complexity have more niches—more different ways to make a living—available for a greater variety of specialized species. For example, coral reefs have a complex physical structure, by virtue of the intricate shapes of the coral itself.
- Patchy or uneven distribution of physical or chemical conditions, such as temperature, salinity, and nutrients, over the area of an ecosystem leads to correspondingly patchy distribution of species and increases the diversity of the whole ecosystem.
- Greater climatic stability (less variation over time) generally results in higher diversity because species do not have to be as adaptable and tolerant of oscillations and sudden changes in environmental conditions.
- When temporal variations in climate do occur, predictable and cyclical patterns like seasonal weather cycles do not restrict diversity as much as irregular instabilities, because species can adapt or specialize to coordinate with the cycles.
- The intermittent removal of patches of organisms by indiscriminate physical factors (e.g., storms or ice) actually increases species diversity by allowing new species to move into an area, unless the frequency is so great or the disturbance so lethal that no species can become established, as happens in areas where commercial trawling for bottom fish is common.
- Fragmentation of ecosystems by physical or chemical interruptions may reduce species diversity and isolate species in smaller areas, making them more vulnerable. For example, chemical pollution may create barriers that interfere with the dispersal of larvae essential for periodic replenishment of animal populations.
- The concentrations of nutrients undoubtedly affect species diversity, but whether they enhance or reduce it depends upon a number of interacting factors. Generally speaking, small intermittent influxes

of nutrients tend to increase species diversity, while large or constant influxes tend to reduce it.

Biological Factors Affecting Species Diversity

- High rates of biological production in marine ecosystems may be associated with either greater or lesser species diversity. Generally speaking, medium levels and more constant levels of productivity support the greatest species diversity.
- Competition between species may reduce or increase biodiversity, depending on how even the competition is and how many competitors interact. A simple competition between two species that are unevenly matched leads to the elimination of the weaker species, but the outcome may be altered by other competitions or other physical factors that inhibit the stronger species. Competition may also affect predator-prey interactions in ways that often allow the coexistence of a greater number of species.
- Predation affects biodiversity in several ways. It often favors higher species diversity by evening out the competition between prey species when the more populous prey species are consumed at a greater rate than the less populous prey species. Predation may also affect the size distribution of prey species, which in turn may affect how many species can coexist in a limited space.
- The coevolution of assemblages of two or more species that interact with each other generally results in the maintenance of increased species diversity; although the loss of one of the species may precipitate the loss of the other(s), thus compounding any reduction in species diversity.
- Mutualism—any purposeful relationship between individuals of two species that brings mutual benefits—generally increases biodiversity by giving survival advantage to both species. Mutualistic relationships are common in marine ecosystems. Included in this category are symbiotic relationships in which there is an intimate physical association between the organisms, as exemplified by zooxanthellae, single-celled algae that live inside the bodies of coral polyps and enhance their nutrition.
- Trophic structure—the length of food chains and the complexity of food webs—has implications on species diversity. Longer food chains in simple systems, which have more or less linear food dynamics, are usually associated with greater species diversity that short food chains; but shorter food chains in complex ecosystems are associated with greater species diversity, because they reflect more complicated food webs with numerous intersecting short food chains.

- Patchy distribution of species is associated with greater total species diversity over areas large enough to incorporate the patches.
- Endemism increases species diversity on a large or regional scale because patches of endemic species increase the total number of species overall. But on a smaller, local scale, diversity may or may not be enhanced by endemism, and a smaller ecosystem may be more vulnerable to loss of diversity when endemic species disappear. Species that have low dispersal and reproductive rates are more likely to be endemic, restricted to a local region.
- Although human activities are not usually included in a list of this sort, they can, and often do, disrupt ecosystems and cause the loss of species diversity. More and more, the humans are having a significant affect on the balance between factors favoring high diversity and factors that reduce diversity in the sea as well as on land.

Diversity Dynamics in Marine Ecosystems

Some interesting theories about how species diversity is determined and maintained in marine ecosystems have been derived from small-scale manipulative experiments in the field combined with observations and long-term monitoring on larger scales. Two theories based on trophic dynamics permeate the scientific literature and have been especially appealing to marine ecologists for explaining many ecosystems. It is likely that both operate to some extent in most marine ecosystems.

The first is based on top-down dynamics, also called *trophic cascades,* dominated by top-level predators. Some predators have roles that are more important to the structure of the entire ecosystem than their abundance or biomass would suggest. These are known as *keystone species,* and fluctuations in their populations can have significant effects on the entire community in which they live. It is the removal of the keystone predator that reveals its importance. For example, in an ecosystem dominated by living biological structure, a keystone species may prey upon and thus reduce the population of a species that in turn preys upon the primary structural species. Removal of the predator leads to the destruction of the structure, which leaves other species essentially "homeless," causing their disappearance as well. Keystone predators may also exert their influence by controlling competition among prey species so as to prevent the exclusive dominance of one species. Intertidal ecosystems are often dominated by top-down dynamics.[4]

Bottom-up dynamics may also determine species diversity. In this case, variations at lower trophic levels and even variations in nutrient levels have effects that "cascade" upward to higher trophic levels. For example, increases or decreases in abundance and diversity of primary producers and grazers affect the species diversity of the entire community, since they lead to corresponding changes in the number of species of each successive level of predators.

Coastal upwelling systems are generally dominated by bottom-up dynamics. Here, nutrient-rich bottom waters mix into the surface waters, stimulating phytoplankton production and increasing production on up the food chain. The diversity of these systems is maintained by the continual supply of nutrients. If that regime changes, as it does off the western coast of South America during El Niño years, diversity is lost as fish and invertebrates disappear or are diminished and populations of seabirds and sea mammals collapse.[5]

Assessing Real Species Diversity Changes in Ecosystems

Ecologists study the biodiversity of ecosystems by measuring changes in species composition and populations over time. They refer back to a standard stable *equilibrium*—the ideal or unstressed community structure. The reference point or starting point for sequential measurements is called a *baseline,* so assessments of an ecosystem's diversity and stability are relative to conditions in a baseline community.

It is important to keep in mind that equilibrium itself is a relative term. It is normal for ecosystems to undergo changes, both cyclical and noncyclical. Therefore, only long-term monitoring of a healthy ecosystem can provide insight into its equilibrium state, the norm about which the ecosystem fluctuates. Some ecologists have also suggested that there are alternative equilibrium states; that is, there may be more than one community of species able to maintain a stable state in a given ecosystem. Clearly, it is difficult to determine the stable state, and it may be equally difficult or impossible to determine when the limits of stability are near and the system is particularly vulnerable to "collapsing" into a degraded state from which it cannot rapidly rebound.

Shifting Baselines in Marine Ecosystems

There are many marine ecosystems about which little is known, and they are being changed rapidly by the numerous stresses discussed in the previous chapter. Most marine communities have not been monitored for change over long periods of time, and their true baselines are unknown. As new research begins, baseline data are collected for reference, but the measured baseline may not represent the most stable or most biologically diverse state for that ecosystem—the system may already be disturbed and unstable. If the measured baseline is not the true baseline for a particular ecosystem—if we do not know the conditions and diversity characteristic of the equilibrium or pristine community and we end up characterizing an ecosystem in a depleted state—subsequent ecological assessments are skewed, and the guidance scientists give about protecting the ecosystem is less valid. Not only is the science compromised, but our expectation or notion of what the natural ecosystem should look like is distorted. We may unknowingly strive to maintain an ecosystem in a reduced and unstable state instead of trying to return it to its natural condition.[6]

A well-known and respected oceanographer, Paul Dayton, professor of oceanography at Scripps Institution of Oceanography, has deep concerns about shifting baselines. In referring to the sweeping damage fisheries have wrought in the marine environment, he stated in 1998 in *Science* (the prestigious weekly journal published by the American Association for the Advancement of Science):

> One irreparable consequence of this widespread damage is the loss of the opportunity to study and understand intact communities: In most cases there are no descriptions of the pristine habitats. The damage is so pervasive that it may be impossible ever to know or reconstruct the ecosystem. In fact, each succeeding generation of biologists has markedly different expectations of what is natural. . . . As with the loss of human cultures and languages after the passing of the elders with their wisdom, so too is humanity losing the evolutionary wisdom found in intact ecosystems.[7]

Genetic Diversity

Up to this point the discussion has focused on species diversity because it is what ecologists use to characterize biological communities and ecosystems. However, genetic diversity and populations are also important. Moreover, in the marine realm in particular, genetically distinct populations may be as significant as species—or, in some cases, what appears to be a population of one species may actually be made up of several different species.

Population Genetics

Taking advantage of a dynamic environment that is constantly in motion, many marine species have developed great dispersal capacity by producing large numbers of eggs and larvae that drift on ocean currents, allowing them to broadly populate areas that have favorable environmental conditions. Thus, it might be expected that prolific genetic interchange among populations of marine species would lead to a quite homogeneous mixing up of genes. In fact, however, there are significant genetic differences among separate populations of many marine species—a surprising bit of knowledge gleaned from modern biochemical techniques that can identify genetic structure within individual chromosomes. The separation of genetically distinct populations is usually spatial, but for some species there can be a simple temporal separation. For example, one species of microalga was found to have two genetically distinct populations in the same location, one blooming in the spring and the other blooming in the fall.[8]

When there is spatial separation of distinct populations, environmental barriers might restrict genetic flow, as when discontinuities in the environment result in unfavorable conditions for larval survival. Less pronounced environmental differences between centers of adult populations will result in the predominance of different genetic traits. Such separations are particularly likely to happen when the adult stage is sessile (attached to the sea bottom) or at least sedentary (staying close to home despite having the ability to move around). Often, although the separation is not complete enough to cause the different populations to evolve into separate species, it is enough to establish genetically distinct and identifiable populations. An example is the humpback whale, an endangered species that has been found to have genetically distinct populations in different regions of the oceans. At least for some marine plants and animals, the species may not be the most significant unit of biodiversity. Populations may represent important genetic diversity that should not be ignored in measures to conserve biodiversity. To protect a species, it may be equally important to identify and protect separate populations. Furthermore, it may not be advisable to transfer members of one population into another population's ecosystem, as interbreeding could alter the genetics such that the new generation does not fare so well.[9]

Another consequence of genetic variability among populations of the same species is that different populations may respond differently to an environmental stress, such as pollution. In a study of the fourhorn sculpin, a fish used as an indicator species in assessing the effects of pollution on coastal ecosystems where it is found, it was noted that the effects of pollutants on natural populations could not be accurately assessed without concurrently determining whether the populations being compared were genetically equivalent. In highly polluted environments, it is not unusual to find populations of species that are adapted to high levels of pollution. For example, certain species of marine worms have been shown to have a greater resistance to toxic chemicals, such as cadmium, in contaminated mud where they live than do populations living in clean mud. The resistance is passed on to offspring, suggesting a genetic component.[10]

There may be a connection between genetic variability and species functions, as exemplified in species of marine teleosts (fishes with bony skeletons). Specialist species, which have very specific environmental requirements, tend to have high genetic variability among individuals within a population. Generalist species, which have broad environmental tolerances and perform a variety of functions, tend to have less genetic variability within a population.[11]

Sibling Species

Several marine pollution studies, beginning in the 1970s, led to the discovery that some populations originally thought to represent single species actually contain a collection of genetically distinct species that look alike. Such groups of physically similar but genetically distinct species are known as *sibling species.* One of the early revelations came in 1976, from genetic analysis of the marine worm *Capitella capitella,* a species often used as an indicator of pollution. It was discovered that what had been recognized as a single species was actually a genus, *Capitella,* composed of six genetically distinguishable species. Further research identified fifteen separate species. It was concluded that the differences among the species, once recognized, could make the genus an even more sensitive indicator of polluted conditions. But perhaps more important, the discovery was an early clue that there may be many more species in the ocean than originally suspected; the corollary to that proposition is that some marine species may not be as widely distributed as once thought.[12]

More recently, many other sea species, some of them commercially exploited or useful to people in other ways, have been found to be complexes of species. The mussel (*Mytilus edulis*) that has been used for years to monitor pollution is now known to be three separate species, a finding that could account for some of the variations previously attributed to different levels of pollution. The identification of corals may be widely affected by the sibling species phenomenon; for example, a Caribbean Sea coral previously thought to be a single species, *Montastraea annularis,* is now known to be three distinct species and possibly more. Even well-known commercial species of oyster, shrimp, crab, and mackerel have each turned out to be a group of sibling species. Two described deepwater crabs have been shown to be 18 separate species. The common dolphin (*Delphinus delphis*) may not be so common, since genetic analysis has revealed what appears to be two distinct species with different distributions.[13]

In the deep sea, where species are so poorly known anyway, identification is made even more difficult by the common occurrence of complexes of sibling species. Some of the *Capitella* species mentioned earlier are restricted to deep-sea environments. The discovery of numerous complexes of sibling species in the deep sea has altered the early assumption that species there are few and widely distributed.[14]

Diversity of Microorganisms

Microscopic life is very important in all marine ecosystems—far more important than previously thought. It has been known for a long time that microalgae are the essential base of the food web in the sea and that the productivity of marine ecosystems depends on sunlight and nutrients to fuel the growth of these microscopic plants. Fungi and protozoa (microscopic animals) also have been known to be present, though their significance in open-ocean waters is only beginning to be understood. Recent discoveries in both deep and surface waters of the open ocean have revealed an unsuspected density of the tiniest microbes along with the more familiar ones. Among these microbes are the true bacteria, called eubacteria, which are approximately 0.2 micrometers (8 millionths of an inch) in size, and the even smaller primitive photosynthetic prochlorophytes and the bacteria-like, chemo-tolerant archaea. Cyanobacteria (photosynthetic bacteria sometimes known as blue-green algae) and prochlorophytes account for a large part of photosynthesis in the ocean.

As techniques for sampling, microscopy, and molecular genetics have become more sophisticated, allowing analysis of the smallest living particles in the sea, it has become more apparent that bacteria and viruses play an important role in marine ecology and are plentiful at all depths and in all ecosystems of the ocean. It is estimated that there are about 100 million bacterial cells per liter (approximately 10,000 cells per drop) of ocean water, and a similar density of prochlorophytes has been reported in sunlit waters as deep as 100 meters (about 330 feet). On an even smaller scale, viruses have been found in abundance in the open ocean.[15]

Although we now know much more than we once did about the abundance of the smallest microbes, their species have only rarely been distinguished, and we don't know much about how microbial species evolve. Nevertheless, numerous functions have been identified, and they are diverse and important. Bacteria play a major role in the cycling of critical elements such as carbon, nitrogen, and sulfur, and they can be found in all types of environments, with species and assemblages of species performing specialized roles. They digest the complex, carbon-rich organic molecules excreted by microalgae and protozoa, and they accumulate on and break down the organic materials in the feces of zooplankton. In fact, they coat most organic debris in the ocean, breaking it down to release the basic elements for their own nutrition and, in the process, also releasing them into the water. Nutrient recycling thus goes on throughout the depths of ocean water. Some of the larger, denser organic particles and dead bodies of animals sink to the bottom of the ocean, where microbes decompose them. But much of the recycling occurs close to the surface and at intermediate depths, where decomposition of molecules and smaller particles and bodies is completed. Aggregates of bacteria and other particles, both living and nonliving, are seen throughout the sea, constituting what is referred to as "marine snow." It shows up ubiquitously in the background of photographs taken in the deep ocean.

Photosynthesis and chemosynthesis are the biological processes by which elements are combined into the molecules of life. Usually, light is needed to drive the reaction (photosynthesis), but a few special organisms can instead use the energy tied up in simple chemical bonds (chemosynthesis). These processes form the base of the food chain on which all other life relies. Organisms able to synthesize organic materials include photosynthetic prochlorophytes, microalgae, cyanobacteria, and chemosynthetic bacteria. These include many free-living species as well as species that live in mutualistic relationships with certain marine animal species.

The variety of microbial functions of synthesis and decomposition undoubtedly translates into significant species diversity, though most of the smaller species are unknown—probably 99 percent of them. It is clear that great functional diversity is to be found, and it appears that size categories of marine microbes may correspond to functional categories. The size of microplankton species determines their route through the food web—the larger ones are immediately grazed by zooplankton, which in turn fuel the food chain of larger predators; the smaller ones, however, remain part of a microbial loop, capturing and recycling nutrients. These can enter the food web of larger animals either when the microbes aggregate on pieces of organic material or on bodies of larger organisms or when they are eaten by tiny protozoa that are eaten in turn.[16]

Comparison of Biodiversity on Land and in the Ocean

Of the species known to science, about 85 percent (some say more) are terrestrial, but, as already discussed, only a small percentage of species in the ocean are actually known—certainly a much lower percentage than on land. Whether the ocean contains as many species as, or more than, the land remains to be disclosed, but it is clearly a possibility. One thing that is known is that there are more phyla and classes—higher taxonomic levels than species—in the sea. Estimates of these vary somewhat because there are different classification schemes, but of all the phyla of organisms, at least 80 percent include marine species and many fewer include terrestrial species. Of 33 animal phyla in one scheme, 32 are found in the sea but only 12 are found on land; 21 of the phyla are exclusively marine and only 1 is exclusively terrestrial. If all macroscopic organisms (i.e., animals and plants) are considered, 43 phyla are found in the sea and 28 on land. Among animals, 90 percent of the classes (one level below phyla in taxonomic schemes) are marine. Although it is more difficult to compare microscopic organisms because the taxonomy is so incomplete, it has been estimated that there are at least 34 phyla and 83 classes in the sea. It has also been noted that on land, 90 percent of all species are members of one phylum (the one that includes insects and spiders), whereas in the ocean the species diversity is more evenly distributed over numerous phyla.[17]

Some scientists hold on to the belief that only 15 percent of species are marine (some estimate as low as 2 percent); others think there may be almost as many species in the ocean as on land, though most of the diversity is apparently benthic (on the sea floor). Several hypotheses have been proposed to explain the apparent phenomenon that there are more species on land while there are more higher taxa (e.g., phyla) in the sea. They have to do with the course of evolution, differences in dispersion in the two media, the environmental differences relative to magnitude of climatic oscillations and physical structure; differences in grazing dynamics and in the nature of species interactions.[18]

Counting species may, in the end, be an exercise in futility if not folly, as the task is seriously influenced by scientists' better understanding of animals that live close to us and are of a size relatively close to ours, or at least easily visible. Very tiny organisms are elusive. Although new research technologies have allowed us to find many of them for the first time, we cannot study their behavior in nature, and the monumental task of assessing them genetically has barely begun.

Perhaps, as some have suggested, simply counting species is misleading anyway, and functions or "lifestyles" might be more important. A review of the major lifestyles reveals several useful terms for designating the larger categories of organisms by their habits and habitats. These include the following:

- *Pelagic* organisms are those that live in ocean water (not associated with the bottom).
- *Oceanic* organisms are pelagic organisms that live in the open ocean.
- *Neritic* organisms are pelagic organisms that live in coastal water, overlying the continental shelf.
- *Plankton* are organisms that are suspended (they float or are weakly self-propelled) in the water and drift with it as it moves. *Phytoplankton* are plant or other photosynthetic members of the plankton community. *Zooplankton* are animal members of the plankton community.
- *Nekton* are animals that are mobile under their own power or, in other words, are able to swim independently of ocean currents.
- *Benthic* organisms are those that live in association with the ocean bottom. They are collectively called *benthos*.
- *Epifauna* are animals that live in contact with the sea bottom, where they are sedentary or move over the surface or are attached to it.
- *Demersal* animals are those that tend to rest on the sea bottom but swim and feed in the water immediately above it.
- *Infauna* are animals that live within or burrow into the bottom sediments.

- *Meiofauna* are tiny animals that live among the grains of sediment.
- *Intertidal* or *littoral* organisms are those that live on the shore between the highest and lowest tide levels.
- *Sublittoral* organisms are coastal benthic organisms that live deeper than the lowest tide levels.

The fluid medium of the ocean and the variety of benthic habitats open up many potential ways of getting food. There are trophic lifestyles or feeding behaviors that are of major importance in the ocean, but are not found at all on land. As a result, marine food webs tend to be more complex and have more trophic levels than terrestrial webs. Filter feeding—a process of straining the water for food—is one example. It is common among a wide variety of animals from zooplankton to barnacles to baleen whales, and it requires special, often complex, filtering structures. The behavior and mechanics required for filter feeding have obviously evolved because it is an effective way to get food from a liquid matrix. Even humans have copied nature and developed a filter feeding apparatus to glean food from the sea—we call it a fishnet. In view of the functional complexity of life in the sea, if there are in fact fewer marine species than terrestrial ones, it seems there must be fewer marine species with similar trophic roles and, conversely, more terrestrial species with redundant roles.[19]

One factor that may contribute to higher species diversity on land is the presence of large vegetation providing complex physical structure. This is absent in most marine ecosystems, except wetlands, kelp beds, and reefs. The most elaborate structures can be found in coral reefs, where animals (corals composed of colonies of polyps) build an organic structural matrix of calcium carbonate. This matrix leads to increased diversity by increasing the complexity of the physical environment, which in turn increases the number of possible biological interactions. Some of the highest recorded species diversity in the sea exists on coral reefs. However, systems dependent on a destructible matrix may lack resilience if disturbances reduce or eliminate the structure on which the diversity is based.[20]

It is also thought that endemism, the restriction of species to one location or habitat, is much less common in the ocean, where fewer species are more broadly distributed than on land. Marine species, so the story goes, are only rarely limited to small, finite areas—areas usually associated with isolated physical or geochemical structures on the sea floor, such as reefs, trenches, seamounts, and seeps or vents. Even in these environments a surprising majority of species seem not to be endemic. The ocean provides a circulating medium ideal for the dispersal of spores and larvae, so it is not unreasonable to expect species to be widespread. Nevertheless, there is increasing evidence that endemism may not be as rare in the oceans as once thought, at least among the benthic animals.[21]

Nevertheless, it is true that many marine species are widely distributed as a result of their mode of reproduction, which takes advantage of the ability of the fluid ocean medium to broadcast spores and larvae. This has implications for organisms that are attached to the sea floor during the adult phase of life, communities of which are often short-lived and subject to periodic physical disruption. Larval transport and settlement allows these species to reach new habitats and enables established communities to be replenished with larvae from distant locations.

The dispersal on ocean currents of larvae and eggs of pelagic as well as benthic species is another distinction between life in the sea and life on land, and it has significant consequences. It means that organisms in early life stages are separated from adult populations, and therefore parents and young do not associate. This is unlike the situation for most terrestrial animals, whose reproductive strategies result in juveniles being in close proximity to adults and often dependent on them. It also means that in the ocean, a single organism is often dependent on more than one ecosystem during its lifetime. Furthermore, the extreme differences between larval and adult life-forms means that the position of each in the food web and the species with which each interacts may differ significantly.

Physical Properties

Several physical differences between marine and terrestrial environments may play a role in determining species diversity in the two realms. One of the major factors influencing the marine biota is the fluid nature of the oceanic environment, which not only enhances cross-fertilization and dispersal of many marine species but also dissolves and helps to circulate nutrients. Patterns of nutrient distribution in the ocean enhance biological production in some areas and limit it in others, and these nutrient dynamics have a pronounced effect on species diversity. In addition, the circulation of ocean waters often facilitates the broad distribution of toxic pollution, and sea water enhances the reactivity of contaminants, increasing the threat to biological communities.

Marine and terrestrial ecosystems also differ with respect to the magnitude of natural environmental fluctuations. Strong seasonal and interannual fluctuations in the terrestrial climate contrast with moderate fluctuations in the marine environment. The ocean moderates its "climate" by means of its vastness and the great heat capacity of water—a large amount of absorbed heat causes the temperature to rise only slightly, and a large amount of heat must be lost for the temperature to drop slightly. Additionally, because marine organisms are always bathed in water, they avoid the severe stress that land species have to endure during dry periods. Ocean species have not had to evolve ways of getting and conserving water.

The great environmental variability on land requires terrestrial organisms to develop physical or physiological mechanisms to cope with the changes. In contrast, marine species other than those very near the edges of the sea have not had to develop mechanisms to respond to rapid environmental change. As a result of this difference, land species may, in general, be better equipped to adapt to some of the rapid environmental change humans inflict than are species of the ocean.[22]

Terrestrial environments have more pronounced physical boundaries between ecosystems. The fluid nature of the oceans does not preclude the existence of distinct ecosystems, but the boundaries between them are soft and mobile. Except for the very sharp boundary between land and sea, the boundaries between marine ecosystems are defined by currents or strong gradients in physical or chemical properties, such as temperature and salinity, and even light. Although ocean waters are contiguous, currents form subtle boundaries that separate water masses with different properties, and they can form physical boundaries that small organisms don't penetrate. If the currents wander or the gradients become stronger or weaker and migrate, most of the living community moves as well. Ecosystems defined by currents and gradients may be very large in scale and may stay in approximately the same geographic location, though the outer boundaries shift. These are sometimes identified as "large marine ecosystems" for the purposes of research and regulation. There is a predominant latitudinal pattern in the major current systems of the world's ocean, and there is a corresponding biological zonation with characteristic fauna and flora.[23]

Great depth is another property distinguishing the oceanic from the terrestrial environment. This expanded third dimension provides a structure that can support a variety of life-forms not found on land. Species in the ocean tend to be distributed in vertical zones in response to environmental gradients and food supply. Marine plants must remain within sunlit waters, the depths of which vary from ten to a couple of hundred meters, depending on the clarity of the water.

Animal species also separate themselves into different vertical zones according to their feeding habits: herbivores in the upper strata; carnivores at various levels, depending on their particular prey; and detritivores, animals that feed on dead organic matter, in deep waters and on the ocean bottom. Competition among species and pressures of predation may also influence the vertical distribution of species. Thus, depth promotes ecological diversity as well as more complicated food webs, and these in turn mean increased functional diversity and in many cases increased species diversity. The structure provided by seawater itself makes up to some extent for the lack of biological structure that supports diverse communities on land, but it is not as varied.

Just as there is a sharp interface between land and sea, there is a discrete interface zone between the atmosphere and the sea, taking the form of a sea surface microlayer. The microlayer has a character that distinguishes it from the environments on either side. Its thickness has been variously described from as little as 50 micrometers (2×10^{-3} inches) to as much as a millimeter (4×10^{-2} inches).

Functionally, the microlayer is in direct contact with the atmosphere and therefore is the area where chemical substances (including toxins as well as benign substances like carbon dioxide and oxygen) are dissolved into the sea and where bubbles from the ocean release gasses into the atmosphere. It is a cohesive layer of water that resists mixing with the waters beneath—a place where both chemicals and biota concentrate. Many species are found only in the microlayer, clinging tenaciously to the ocean's surface. Other species are present only as eggs, larvae, or floating visitors.

Chemistry and Biochemistry

Because of the omnipresence of water, which can dissolve almost anything, the ocean is a reactive chemical soup that interacts with the biochemistry of the creatures living in it. Animals and plants both on land and in the ocean produce chemicals that send messages to others. They may attract others of their kind to ensure reproduction; they may send warnings to potential predators; they may be harmful to potential competitors and keep them at a distance; or they may promote the growth of other species or of their own kind. One such message that is exclusive to marine environments is the message to larvae that "this is a good place to settle down and metamorphose." The chemicals relaying this message may be produced by adults of the same species or by some other species. In the latter case, the target species (the one that receives the message) will have evolved a sensitivity to a chemical product exuded by another species that indicates a favorable habitat. Such interdependent evolution of species commonly living together is called coevolution.

The larvae of abalone (a commercially harvested mollusk) require a substance produced by a particular species of red alga that is common on rocks in areas where abalone grow successfully. If the red alga is not present and producing its exudate, the larvae will not settle and metamorphose to grow into adults. This mechanism guarantees that the abalone larvae will keep drifting until a suitable environment is found—but it all has to happen within a relatively short period of time, generally a few weeks, if the larvae are to survive. The alga species, which grows as a thin crust on the rocks, also benefits from this coevolved relationship, when the abalone grazes on taller algae that might outcompete the lowly crust not easily grazed by the mollusk. Thus, the two species help maintain each other. Similar relationships are known to exist between the sea hare (an invertebrate) and a red alga and between oysters and certain bacteria. Such chemical interdependencies between species may be quite common in the ocean.

Chemical signals for self-regulation are also common among marine species. For example, mass spawning in beds of abalone is initiated by a hormone released into the water by a few individuals. In nature, mass spawning ensures the maximum number of encounters between released sperm and eggs. In another example of chemical self-regulation, the whelk, a sea snail common in the North Sea, relies on chemoreception akin to our sense of smell to locate mates and prey. There is justifiable concern about the potential for chemical pollutants to interfere with the production, dissemination, and reception of chemical messages. In the case of abalone, for instance, irritant chemicals—in particular peroxide—can stimulate production of the spawning hormone when the animals are not ready. In the case of the whelk, it has been shown that chemical pollutants from dumped waste can render chemoreception ineffective in locating food.[24]

Other pollutant chemistry in the ocean has resulted in the broad dispersal of many persistent toxins and their ultimate concentration in polar regions. A category of persistent pollutants called organochlorines (including PCBs and DDT, for instance) are among the most infamous and ubiquitous human-made toxins. They are not broken down in the environment and thus persist and accumulate in seawater, sediments, and living tissue. Both PCBs and DDT are carried to the ocean in runoff from land and may also pollute the atmosphere and fall onto the ocean's surface. The alternating volatilization and condensation of these substances, combined with ocean and air current patterns, has carried them from their points of origin in lower latitudes toward the poles, especially into the Arctic. Consequently, the water, sediments, and biota—including humans—of the Arctic contain exceptionally high concentrations of organochlorines, even though they are not used or manufactured there. The people are threatened with severe health problems because of this pattern.

The ocean is also a place where natural toxins are produced. Many marine animals, particularly in tropical reef areas, produce a variety of toxins to ward off predators. Even more ubiquitous are the toxins produced by numerous species of phytoplankton. These metabolites are excreted into the water in a type of microscopic warfare designed to reduce the competition. In normal, diverse plankton communities, effects of the toxins remain at the microscopic level. However, if one of the toxin-producing species finds just the right growing conditions and gains the upper hand, it can bloom to densities that release enough toxin to poison larger animals, such as fish, whales, or humans, that consume or contact either the microalgae themselves or some animal that has consumed them or their poison. Such occurrences, often known as red tides or more generally as harmful algal blooms, are increasing in frequency. This is due at least in part to the increase in nutrients flowing into coastal waters from centers of human population and agriculture. There is no terrestrial equivalent to red tides.

Scales and Patterns of Biodiversity Distribution

Biodiversity in the ocean is determined by the interaction of numerous processes, physical and biological, that operate on various spatial and temporal scales. Furthermore, ecosystems themselves can be defined on various spatial scales. When a particular ecosystem is identified for the study and/or conservation of biodiversity, it is important to recognize the interaction of processes on different scales that can affect diversity in the system over time.

Although global biodiversity is of ultimate concern for the future of our living planet, rarely do scientists, policy makers, or regulatory authorities address biodiversity on this scale. Local and regional scales are far more understandable and manageable, for the ocean as well as for land. Scientists define these smaller scales on the basis of physical and biological processes, and regulators and policy makers define them on the basis of jurisdictions. The ocean, of course, does not recognize political boundaries, so more effective management schemes will take into account the scientifically defined boundaries of ecosystems. Even then, protection of biodiversity is difficult because of the open nature of marine ecosystems, which readily exchange their biological and chemical constituents over space and time. For instance, if the species diversity of a particular estuary is of concern, it is necessary to look beyond the estuary to neighboring coastal ecosystems that may exchange biota with the estuary; other estuaries in the region that may be sources for species replenishment; rates of water exchange; sources of pollution (some of these may be very far away); annual weather patterns; and climate trends over time. If a marine ecosystem on the open coast or in the open ocean is the focus of attention, additional processes become important, including currents, upwelling, and patterns of turbulence, which may vary over both space and time. All processes mentioned so far operate on scales that reach beyond the confines of ecosystem boundaries. Within the ecosystem of concern, physical and biological processes acting on smaller scales can give rise to patchiness in the distribution of species. Patches have their own species diversity but they also increase diversity in the broader ecosystem.

Gradients of Species Diversity

On the basis of fossil data as well as current knowledge of life in the sea, several widespread patterns of species distribution have been identified. However, they vary somewhat from one type of organism to another, and even then exceptions are always cropping up. Perhaps the most inclusive studies are those of benthic animals in the deep ocean. Because of all the exceptions and because marine biologists studying extant species rarely look at more than one group at a time (e.g., seaweeds *or* zooplankton *or* reef fishes *or* corals—and, not surprisingly, patterns vary), these gradients are only briefly mentioned here.

One of the most common patterns corresponds to latitude. An increase in species diversity with decreasing latitude has been reported for some fossil groups and for living mollusks, and there is some indication that diversity associated with deep-sea vents may have a similar pattern. Different patterns have been found in planktonic animals, for which researchers have variously reported diversity peaks at fifteen degrees and at midlatitudes around forty degrees. Seaweeds also have a species diversity peak at midlatitudes. A longitudinal diversity gradient has been noted for coral reefs, with a maximum species diversity in the eastern Pacific Ocean. Pacific coral reefs are, on the whole, more diverse than Atlantic reefs, and in both oceans there is an eastward trend of decreasing diversity.[25]

Another is the gradient in species diversity in an offshore direction from continental margins to the open ocean. It was long thought that the diversity of benthic species decreased with depth, but more recent studies, in which smaller species were sampled, have revealed just the opposite pattern, perhaps with a peak at intermediate depths around 1,500–2,000 meters (5,000–6,700 feet). The diversity of open-ocean pelagic species, however, does not seem to follow this pattern, and estimates suggest that the total diversity of open-ocean waters is comparable to the diversity found in the narrow band of coastal waters.[26]

Although there are no clear geographic patterns that predetermine the species diversity of an ecosystem, there is probably a limit to the potential species diversity in a region. Just how much of that potential is realized depends on the interaction of physical and biological processes acting on the regional pool of species.

Carving up the Ocean

The ocean is a big place, so, quite naturally, scientists and policy makers prefer to divide it into pieces when studying it or regulating activities that affect it. The largest subdivisions have to do with the most basic nature of the habitat—is it mostly solid or mostly liquid? Thus, there are the benthic and the pelagic marine environments—a division based on both physical and biological properties. The benthic, or bottom, habitat supports a very different kind of animal and (sometimes) plant community from that found in the fluid body of the ocean.

Further dissection of the pelagic and benthic realms divides the ocean into several more vertical and horizontal regions. Vertically, the benthic realm includes the intertidal shores, the shallow continental shelf breaking at about 200 meters (660 feet) into the continental slope, which is sharply inclined down to the abyssal plane, which spans most of the deep sea at depths of 3,500–4,000 meters (about 11,500–13,000 feet). The sea floor is occasionally cut by trenches, the sides and floors of which are called the hadobenthic zone, and may extend as deep as 11,000 meters (or nearly seven miles). The pelagic realm is stratified from top to bottom into the photic or epipelagic (above 200 meters), and the aphotic zone (below 200 meters), which includes mesopelagic (down to 1,000 meters, or about 3,300 feet), bathypelagic (down to 3,500 meters or about 11,500 feet), abyssopelagic (down to 6,000 meters or about 19,700 feet), and hadopelagic (down to 11,000 meters) zones.[27]

Horizontally, the ocean is classified into coastal and oceanic regions. Coastal ecosystems are relatively shallow and characterized by important interactions with the land. Coastal waters are most affected by human activities, and as a result most coastal ecosystems have been changed to some degree. The inner boundary of the coastal region is the continental shoreline; the outer boundary, which defines the transition between coastal and oceanic regions, is determined by the ocean bottom topography and lies at the edge of the continental shelf. The distance from shore varies considerably from one place to another. Because of their interaction with land, coastal ecosystems are more nutrient rich than the open ocean. The distinction between coastal and oceanic regions or zones has more political significance than do vertical zones, since it corresponds to the demarcation of Exclusive Economic Zones, usually set at 200 miles offshore, inside which the bordering country has jurisdiction over the resources and outside which all resources are international. Nevertheless, it is important to note that the ecological boundary between coastal and oceanic is soft, and there is physical and biological communication between the two. Many species are restricted to one side or the other, but others cross between the two realms. Understanding their interconnectedness is crucial to understanding ocean ecology and to protecting ocean biodiversity.

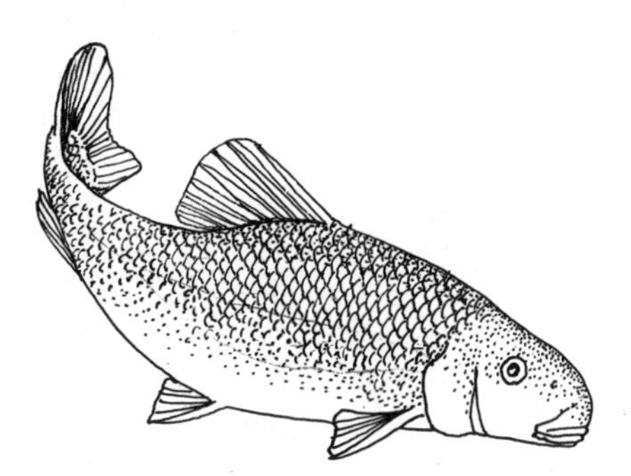

Chapter Four

Coastal Marine Ecosystems

The division of the ocean into its various physical regions, as discussed in Chapter 3, is convenient for research and regulation, but the marine environment is first and foremost an assemblage of interconnected and interdependent ecosystems. It is often difficult to define one marine habitat except in the context of others, especially when discussing animals that spend different stages of life in different habitats.

Nowhere is this interconnectedness more consequential than in the coastal zone, where the continuity does not stop at the end of the sea but includes critical interactions with the land and the atmosphere as well. As much a concept as a place, the *coastal zone* has been defined as extending at least from the inland limit of tidal or sea-spray influence to the outer extent of the continental shelf. Estuaries and coastal river mouths are included because they are influenced by tides. Some scientists prefer to extend the inland boundary of the coastal zone into river basins that drain into the sea because their freshwater outflow and their loads of sediment and dissolved materials make them an integral part of coastal zone dynamics.

Although it occupies far less area than either the continental land masses or the open ocean, the coastal zone is a critical area of chemical and biological exchange. The total area of coastal zones globally is about 10 percent of the area covered by the open ocean and a little more than 20 percent of the area of dry land. The coastal zone is the most productive part of the ocean, and therefore it is important in the global carbon cycle as well as in the global cycles of other elements and water. It is also an area of both species segregation and integration, since species may sort themselves along physical gradients, but some cross these gradients at different stages in their life cycles. It is, as well, subject to the greatest accumulation and synergy of insults from human activities. Human-generated nutrients, toxic pollutants, and pathogens wash from the land, are deposited from the atmosphere, and are discharged from ships and oil rigs; hunters, fishers, gatherers, and ocean farmers sweep wildlife from coastal waters, shores and sea floors; freshwater is captured for irrigation and drinking water, which reduces the natural flow into the coastal zone by as much as 50 percent; and human populations and development encroach on marine habitats.

Within the coastal zone, there are several types of marine ecosystems: estuaries and wetlands, rocky and sandy shores, coral reefs, benthic ecosystems of the continental shelf, and coastal pelagic ecosystems. Each type of ecosystem is introduced in this chapter with a discussion of its ecological processes and biodiversity and the role of humans in changing the ecosystem.

Estuaries and Wetlands

Estuaries and associated wetlands lie at the interface of the marine environment and the land. This boundary is characterized by bays and sounds, river mouths, lagoons, and inlets—estuaries, where inflowing freshwater is at play with ocean tides, creating daily variable salinity gradients and a variety of habitats.

Because of fluctuating sea levels and movements of the earth's crust resulting in spreading sea floors and continental drift, continental margins have been redefined many times throughout geologic history, so estuaries are generally very young on this time scale. As a consequence, they have not had time to evolve complex communities with high species diversity. As beautiful and diverse as they seem to us, their diversity is relatively low compared with that of more deeply submerged marine ecosystems. However, because estuaries are fairly well defined by their shorelines and are separated from one another by intervening land and open ocean, the communication between estuaries is limited and they may have very different biotas. Thus, collectively they represent a greater species diversity than is associated with each one alone. One consequence of its early evolutionary stage is that an estuary may have numerous uncast roles awaiting new actors to try out for them, so when foreign species (from some other estuary) are introduced, as they frequently are by shipping activities, they may find a home that they fit into well or even one they can dominate at the expense of species already there. Copious accidental species introductions have resulted from human activities and are now recognized as one of the principal scourges of estuaries around the world. It is impossible to predict which transfer will cause devastation to a natural community, which will merely augment it, and which will fail.[1]

Daily and seasonal variability dominates an estuary's environment, so physical and chemical factors are of greatest importance in controlling its biology. Salinity gradients—from freshwater to full-strength seawater—and seasonal changes in temperature, light, and nutrients define the habitat and regulate the distribution of species in it. The large salinity gradients fluctuate predictably with the tides and the seasons and with distance from freshwater inflows and open coastal seas. There are also irregular and unpredictable changes in the salinity regime during storms, which may increase freshwater flow from rivers or saltwater intrusion from the ocean. The inconsistency of the physical environment may contribute to the relatively low species diversity that characterizes estuaries.[2]

Despite these fluctuations and inconsistencies, the general gradient of salinity, from near-seawater at the outer reach to near-freshwater at the inland reach, dictates a corresponding gradient of species diversity that declines progressively from the coastal sea into less saline, or brackish, waters. The decline is primarily due to the loss of whole taxonomic families of species as the salinity decreases. However, in the freshwater of the rivers, species diversity again increases. A few tropical estuaries actually have salinities greater than that of ocean water, caused by an excess of evaporation over rainfall, and these tend to harbor lower species diversities.

Nutrient dynamics play a critical role in the productivity and species diversity of estuarine systems. Most estuaries have a seasonality related to rainfall and wind, which wash nutrients from the land and stir up nutrient-rich bottom sediments. At mid- to high latitudes, seasons are defined by differences in temperature and sunlight as well. A combination of increased nutrients and sufficient sunlight causes a growth spurt in microalgae as well as the larger algae and submerged vegetation, and this in turn fuels increased growth and reproduction of animals. There is a corresponding seasonality in the species composition of microalgae, which rest as dormant spores on the estuary floor during periods less favorable to their growth, and there is some variation in the time when each species experiences its peak growth. Animal species include those that live in the estuary year-round, those that migrate through the estuary from river to sea or from sea to river, and those that periodically move into the estuary from the sea to reproduce and/or feed.

Estuaries function as nursery grounds for many species of coastal and oceanic fish. Consequently, environmental conditions that affect species as larvae and juveniles in an estuary can greatly affect the populations and distribution of adults in other habitats. In the United States, an estimated three-quarters of the fish species in East Coast estuaries spend part of their lives in the coastal waters outside the estuaries, so conditions within the estuaries have a direct effect on the fish communities outside. In fact, some of the productivity of outer coastal habitats is exported directly from estuaries. A lesser transfer of biological production moves in the opposite direction.[3]

Estuaries are fringed and dotted with wetlands of various sorts. Salt marshes in the temperate zone and mangrove forests in the tropics grow in the brackish intertidal areas of estuaries. Seagrass meadows, usually growing entirely submerged (called SAV, submerged aquatic vegetation), are common in both climes, although they are characterized by different species of seagrass. Salt marshes and seagrass meadows are dominated by species of flowering plants, which provide physical structure for the habitat and harbor numerous species of seaweeds and small animals, including the young of open-water species. There may be hundreds of species of drifting algae associated with them, but these do not support as diverse a fauna eating them. The flowering plants contribute to the productivity of the estuary and the coastal waters beyond it, primarily through a food chain involving organisms of decay and detritivores (animals that eat rotting organic material).[4]

Mangrove forests once occupied about three-quarters of tropical protected shores and inlets, creating estuarine habitats that harbored a distinct diversity of associated species. Approximately fifty species of salt-tolerant woody plants, known generally as mangroves, are distributed along a diversity gradient, similar to the gradient known for coral reefs, that decreases from west to east, with a peak in the western Pacific Ocean where it meets the Indian Ocean. Numerous plant and animal species are associated exclusively with these ecosystems, and others move in and out of them. The mangroves' underwater but aboveground root systems form a natural protected area for small estuarine animals and the young of other coastal species. These ecosystems are severely threatened by human activities—the expansive aquaculture of shrimp in ponds built by cutting down mangrove forests, deforestation from firewood collection, conversion to farming, tourism, and urban development. [5]

Intertidal mudflats, unlike other types of wetlands, are not dominated by vegetation. These are sites where animals burrow into the mud, providing food for numerous coastal birds. The mud-dwelling species—which are not great in number—are predominantly worms, clams, snails, and a few noteworthy crustaceans such as fiddler crabs. These curious little asymmetric creatures, with one very large foreclaw and one small one, are ubiquitous on mudflats and are symbolic of life there. They have a behavior especially adapted to the alternating submerged and exposed conditions caused by the tides. They build individual airtight burrows, with plugged entrances, and sit out high tide. Then, at low tide, they all open their doors and scurry out onto the mudflat to feed on small organisms in the mud and organic material left by the tide.[6]

In addition to macrophytes, the large plant types already mentioned, microalgae are an important part of an estuary's primary production. As estuaries become increasingly nutrient rich by receiving runoff and atmospheric inputs of nitrogen and phosphorus, microalgae account for an increasing amount of primary production, and seaweeds and seagrasses become diminished in their area of coverage, biomass, and diversity. It has been estimated that nitrogen influx into coastal estuaries in the eastern United States has increased approximately tenfold since prehistoric times.[7]

The conversion from submerged macrophytes to microalgae is further promoted when populations of important filter feeders that eat microalgae are reduced by disease and/or overfishing. A severe example of this is seen in Chesapeake Bay, where oysters once filtered the entire volume of the bay about once a day. Now the population is so diminished that it barely filters that much water in a year. First, oysters were dredged out and overexploited, with no attention paid to the simple necessity of returning the empty shells to the estuary so that oyster larvae could settle on them and develop new populations. On top of that, the oysters have been plagued by two protozoan diseases. Fertilizer runoff, air pollution deposition, and reduced grazing have resulted in densities of microalgae that make the deeper water nearly devoid of light; thus, in many places in the bay, submerged vegetation can no longer grow. Yet another compounding factor has been continued degradation by wasting disease of the predominant seagrass species, *Zostera marina,* since the first massive outbreak in the 1930s. This is just one of many hundreds of stories of estuaries around the world being changed and degraded by a multitude of negative events, most of them traceable to excesses in human activities.[8]

Two other stories have possible ties to Chesapeake Bay—one involves an estuary in Washington State, and the other an inland sea in eastern Europe. The oyster species native to the Chesapeake and other East Coast estuaries, *Crassostrea virginica,* was imported into Willapa Bay in Washington for the purpose of establishing harvestable populations—a kind of free-ranging aquaculture. With no thought of possible consequences, oyster spat (young animals) from the East Coast were wrapped in a local species of cordgrass, *Spartina alterniflora,* the dominant salt marsh grass on the East Coast. Now, a century later, that grass is taking over the important mudflats of Willapa Bay. Washington State officials feel compelled to take drastic and controversial measures to try to eradicate it, even though important habitat may be lost and other species damaged in the process, and it is unlikely that the offending species can be eradicated at this late stage. In the meantime, even though the East Coast oyster didn't work out, another alien species, the Pacific oyster from Japan, has been successfully introduced, with the blessing of the state, as the basis of a major oyster industry. The biology of Willapa Bay has been significantly modified by the growing and harvesting of oysters and the measures taken to prevent other species from interacting negatively with the imported species. One apparent victim is a small native oyster, *Ostrea lurida,* which has all but disappeared from the bay. A different set of circumstances has carried an accidental tourist, most likely from Chesapeake Bay, into the

Black Sea of eastern Europe, where it has wreaked havoc. Transported in ballast water, the American comb jelly (*Mnemiopsis leidyi*) was so successful in its new home that it consumed a large portion of the native plankton in the Black Sea. It outcompeted many of the native animals that feed on small plankton, and comb jellies are not a suitable food for native predators of the displaced fauna. There was a 90 percent decline in plankton biomass, and the resulting overall loss of biodiversity throughout the food web was devastating.[9]

Horseshoe crabs, are intriguing and important estuarine animals represented by four species worldwide (*Limulus polyphemus* on the East Coast of the United States). They are also seriously threatened by human activities. Often referred to as "living fossils" because they have survived for hundreds of millions of years, unharmed by conditions that brought down the dinosaurs and countless other species, horseshoe crabs now are in trouble—threatened by a multitude of human activities in the estuaries where they live and reproduce. They are arachnids, more closely related to spiders than to crabs and even more closely related to animals long since extinct. They also play a vital role in the estuaries where they live. The East Coast horseshoe crab, for example, is critical to the shorebirds that fuel their spring migration each year with a stop at Delaware and Chesapeake Bays to gorge on horseshoe crab eggs, which are deposited in sand along the shores. This feeding frenzy coincides with the horseshoe crab's mass spawning during the highest spring tides as the water warms in May and June. But these creatures are severely threatened by overfishing (they are collected for fertilizer, fish bait, and medical research) and pollution. During the 1990s, numbers of spawning adults on Delaware's beaches declined as much as 90 percent, a loss that has been attributed to increased take as bait for the area's growing eel fishery.[10]

Human effects on estuaries seem endless. Among the most direct and permanent effects is the outright destruction caused by filling in wetlands, dredging, and construction of roads, ports, marinas, farms, and other structures that encroach on the estuarine environment. Such activities have significantly reduced the area of many an estuary, often destroying a significant portion of associated wetlands. Estuarine areas have also been altered by the construction of dams and the diversion of river water to be used for irrigation and human consumption. The reduction of inflowing water causes significant loss of wetlands and changes in salinity regimes of estuaries, typically resulting in a loss of species diversity.

Another serious problem is toxic pollution of estuary sediments caused by industrial and sewage discharges, runoff from land, and airborne emissions that fall on the waters. Most wetland sediments are naturally anoxic (lacking in oxygen), so animals cannot live deep within them, although some flowering plants specially adapted to wetland environments can survive there. Wetlands are also known and valued for their ability to filter toxic substances from the water and store them in the sediments, which settle out when the rushing inflow of rivers slows down and spreads out over the larger area of the estuary. Estuary sediments are rich in decayed biological or organic matter, some produced within the estuary and some washed from land, which can accumulate very high concentrations of toxic chemicals that preferentially adhere to organic material. These warehouses of toxins threaten the life-forms that live and feed on the estuary bottom.[11]

Nutrient pollution also threatens estuaries even though they are areas naturally rich in nutrients. The very nature of many estuaries has been changed by nutrient inflow from agricultural and sewage runoff and deposition of air pollutants. Waters once clear and dominated by submerged large vegetation are now highly turbid and dominated by excessive growth of planktonic microalgae. Nutrification often becomes so intense that harmful algal blooms persist for long periods of time. Besides threatening human health, such blooms may diminish the diversity and productivity of the natural estuarine community, at least for a period of time.

One of the most startling examples of such a bloom occurred repeatedly in the 1990s along the middle Atlantic coast of the United States, primarily in brackish-water rivers of North Carolina, Maryland, and Virginia. Identified as *Pfeisteria piscida*, the tiny one-celled creature—part plant, part animal—caused large-scale devastation, including the destruction of millions of fish and some serious health effects for people who merely touched the water. Seemingly out of a science fiction novel, this creature kills fish with one or more potent toxins and then feeds on the dead tissue. Its proliferation has been linked to excessive nutrient runoff from wastes generated by intensive development of industrial-scale animal farms raising hogs and poultry.[12]

One of the important issues regarding estuaries and wetlands is that of restoration. Can the adverse effects of human activities on these sensitive ecosystems be reversed by efforts to restore them to their characteristic biodiversity? Once damage has been done, simply ending the offending activities may not be enough to restore these systems to their original state.

Rocky and Sandy Shores

Marine ecologists have carried out extensive studies of rocky intertidal and shallow subtidal ecosystems, and from these studies have emerged some basic theories about the ways in which species diversity is regulated in marine ecosystems. Intertidal and subtidal ecosystems are relatively accessible to researchers, and their exceptionally interesting flora and fauna fascinate naturalists around the world. Their diversity of species is moderate to high, in part because of their physical structure, which can provide a great variety of living conditions. Their biotic communities are made up primarily of seaweeds and invertebrate animals that are attached or cling to the rocks, and in some areas marine mammals may have an important role. Solid rock may be sculpted or boulders irregularly arranged, thus furnishing nooks and crannies to hide in, rock faces at various angles to the waves and currents, and protected crevices and pools.

Space is the principal commodity for which plants and animals compete, and the outcome of these competitive interactions is governed by a variety of physical and biological factors. The oscillation of the tide provides a dynamic environment: food and nutrient supplies are refreshed with each incoming tide, and reproductive spores, larvae, and organic material are dispersed on outgoing tides and longshore currents, which are currents that run parallel to the coastline.

The fluctuating tides leave intertidal areas—the areas between the highest and lowest tides—variously exposed, resulting in a clear zonation of species; the zones can usually be distinguished as different-colored bands reflecting the colors of dominant species. The assemblages of species in each zone represent the winners of the competition for space under different regimes of low-tide exposure. More species are present in lower intertidal and shallow subtidal zones, where the physical environment is less stressful, than in higher zones, where periods of exposure are the longest. Seaweeds flourish in the rocky intertidal regions, and major kelp beds line shallow subtidal shores. Rocky shores in temperate regions of the world house the zenith of seaweed diversity. Here, seaweeds thrive because the environment provides a hard substrate for attachment, a variety of niches, enough light, and abundant nutrients carried in the waves of each incoming tide.

Species composition and distribution in intertidal zones are determined by several physical and biological factors: tidal range, desiccation (drying), light intensity, tidal and seasonal temperature fluctuations, nutrients, sources of larvae for recolonization, intensity of predation, and competition. The relative importance of these factors varies from one location to another. Generally speaking, biological factors seem to be the predominant processes affecting diversity and distribution in the lower range of the intertidal zone and in the subtidal zone, whereas physical factors have a predominant effect in the upper reaches of the intertidal zone, where physical conditions are harsh.[13]

Keystone predators (discussed in Chapter 3) seem to be important in many intertidal systems, where research led to the development of the keystone predator theory. The efficiency with which they prevent monopolization of space by competitively dominant species has been demonstrated in field experiments in which the keystone species is excluded from the study area. Comparisons of subtropical and temperate rocky intertidal habitats reveal a higher ratio of top predator species to total species in the subtropical communities, which also have greater total species richness.[14]

Keystone predator species have been identified in various rocky intertidal systems, for example the sea otter in kelp ecosystems. Sea otters feed heavily on sea urchins and abalone, both shellfish that graze on kelp. The sea otters thus keep down the populations of grazers, allowing healthy stands of kelp to continue to flourish. The kelp provides the ecosystem's primary habitat structure, which can support a high species diversity. In areas where the sea otter has been hunted out, sea urchins may proliferate to the detriment and eventual exclusion of the kelp and other species that rely on it. When this occurs, sometimes only a pavement of urchins and little else remains; at other times, the kelp may remain, but the system is impoverished—a mere ghost of the rich community it once was. When the sea otter was hunted for its fur, many areas of extensive kelp beds, from Alaska's rocky shores to the majestic kelp forests off the California coast, were damaged or destroyed. Nevertheless, California's giant kelp forests, seemingly governed by storms and upwelling and at times able to coexist with dense populations of urchins, may be able to survive in the absence of otters. In kelp forests off the coast of Chile, sea otters are absent and sea urchins apparently do not have the same role, so factors determining diversity may differ from location to location in the same ecosystem type.[15]

Another keystone species in rocky interidal ecosystems is a species of sea star preying on mussels, which allows the coexistence of other species that compete with mussels for space on the rocks. If the sea stars are removed, other species disappear as mussels take over their space. The removal of this one species has a cascading effect, leading to the reduction of species throughout the ecosystem. At the same time, though, sea stars may reduce species richness by overgrazing, so the relationship is complicated. Still, there is little question that in many intertidal ecosystems, sea stars are decisive in regulating the diversity of the biological community. Similar keystone roles have been attributed to lobsters along the Atlantic coast of North America and whelks in Chile. Once a keystone predator disappears from an ecosystem, the ensuing changes in the ecosystem may be so fundamental that reintroduction of the missing predator will not return the ecosystem to its former state because it can no longer perform that keystone role. For example, in one study in which sea stars were removed from a community for a period of time, the mussels took over the area and grew too large for sea stars to eat when they were returned. Thus, replacing them had no effect on the species diversity.[16]

Several other biological factors may affect species diversity of rocky intertidal habitats. Competitive interactions in the absence of keystone predators play out in ways that may either reduce or increase species richness. Also, since many intertidal invertebrates are relatively short-lived or subject to removal by physical and biological forces, their replacement by larvae—called recruitment—is critical to their maintaining a position in the community. Thus, dispersal dynamics, as well as reproductive schedules and distances between populations, affect species diversity. The larvae of invertebrates often travel significant distances, so the source of larvae for a given area may be another population some distance away, requiring the larvae to survive the intervening environmental conditions. There is significant predation on larvae, and they may fall victim to unfavorable water conditions such as pollution. Consequently, recruitment is variable from one year to the next.[17] Seaweed populations also must be replenished, but this is generally carried out on a more local scale by spores or drifting pieces of seaweed.

Physical disturbances likewise play a major role in regulating a seashore's species diversity. Periodic damage by storms or scraping by logs or ice clears spaces that can be invaded by species that don't do well in long-term competitive interactions. Generally, an intermediate frequency and intensity of such disturbances favors the highest diversity. With too little disturbance, competition can reach its conclusion, with the weaker species driven out; whereas with continual disturbance, many species simply cannot survive to reproduce and so are lost.[18]

One area of intense physical disruption is the surf zone on rocky shores. Heavy surf action causes organisms to be battered by waves, generally not a favorable situation for species trying to cling to the rocks. Seaweeds can be torn up, and only the most tenacious animals avoid being washed away or tumbled along the rocky substrate. As expected, the diversity of species here is low; however, those that remain are unusually productive, making the surf zone among the most productive marine habitats. Wave action indirectly enhances productivity by continually carrying nutrients and suspended food particles to plants and animals attached to the rocks and by keeping seaweeds wet during low tide so that photosynthesis can continue during the period of greatest light intensity. The seaweeds and invertebrates living in the surf zone are especially adapted to the harsh physical conditions there, but they are not good competitors in quieter waters. Therefore, most are found only in this zone. One particularly intriguing species is the sea palm, a seaweed that cements itself securely to the rocks. It has a thick, flexible stemlike stipe (stalk) that bends with the waves, along with numerous thick, narrow fronds arranged like palm leaves, suitable for absorbing nutrients but resistant to shredding. From a distance, a cluster of these seaweeds on the rocks looks like a grove of palm trees, though they are actually only one or two feet high.[19]

Species diversity in rocky shore ecosystems does not necessarily follow a typical latitudinal gradient with more species found in the tropics. For instance, seaweeds reach maximum density in temperate latitudes, and invertebrates seem to have as high a diversity in some temperate regions as they do in tropical areas. Although a very high diversity of species is known from these systems, it may be because the vast majority of species there are easy to see and to reach and therefore have been identified. The same is not true for sandy shores, which are thought to have a significantly lower species diversity than rocky shores, in part because the substrate is unstable and food sources are limited. However, the *meiofauna* of sandy shores—the tiny animals that live among the sand grains—are not well assessed, so it is not really known how many such species exist.

The species that live on sandy shores are adapted in various ways to their unstable environment. Many burrow at least partway into the sand and usually can retreat even farther into the sand at low tide. Others, such as certain fish and snails, migrate shoreward and out again with the tides. As on rocky shores, there is a zonation of species along the tidal gradient, but it is not as distinct. There is a progressive increase in species from the high-tide zone to the subtidal zone, related to exposure to drying and wave action. The tiny meiofauna living among sand grains are an important component of the community, serving as a source of food for some of the larger animals living on or in the sand. Although the meiofaunal species are poorly known, they may outnumber the species of larger animals. These tiny creatures exhibit the typical tidal zonation pattern and also show zonation with depth in the sand relative to wetness, temperature, and oxygen level. The base of the food chain on sandy shores is primarily microalgae that wash in with the tides. Along some beaches, a few highly productive species of diatoms—types of microalgae—live exclusively in the surf itself, and masses of them are left on the surface of the sand during low tide. Filter feeders consume them when they are suspended in the water, and animals that burrow and feed in intertidal sand, such as razor clams, feast on them during low tide.[20]

For the same reasons that rocky intertidal and subtidal areas are attractive to scientists, they are subject to a variety of abuses resulting from human activities. Overfishing, polluted runoff and waste outfall from coastal development, and habitat destruction or modification commonly plague these ecosystems. In some places along the coast, even scientific research and educational field trips have taken their toll, robbing the communities of many of their natural inhabitants when they are collected by well-meaning students and researchers. The tides and waves bring in protective water, laden with food, but often they are also laced with harmful heavy metals, petroleum products, toxic organic chemicals, and litter. Some seaweeds and animals slip off the rocks; some become diseased or deformed or die; some cannot reproduce successfully. Larvae may not find the proper environment or the requisite chemical signal to make them settle successfully, so aging populations are not adequately replenished. Eventually, some species disappear and others lose genetic variation within their populations. In short, biological diversity is diminished. The enormity of the transformation of intertidal communities in Chile by people gathering shellfish, fish, and seaweeds was realized only after certain parts of the coast were closed to human access and, in the absence of human pressure, a much different and more diverse intertidal biota arose.[21]

Coral Reefs

Coral reefs have the greatest known species diversity of any marine ecosystem, although it has been suggested that the deep-sea floor might have as great or greater species richness. Many people are more familiar with the diversity of coral reefs than with that of other parts of the sea because of their relative accessibility, the increasing popularity of scuba diving, and the inherent beauty and photographic appeal of their colorful biological communities. Crisp photographs are possible in clear tropical waters, and the variety of closely packed colors, forms, and motion inspires awe in divers and armchair divers alike. The richness of animal species is similar to that in tropical rain forests, but the plant species are considerably fewer. Both ecosystems have a biological structure that provides numerous niches for different species to fill; but, whereas in the rain forest the structure is provided by plants, in the coral reef it is provided by animals. Corals, which are colonies of individual animals called polyps, build elaborately formed calcium carbonate "condominiums."

Despite the familiarity of coral reefs to scientists, tourists, and television viewers, it is estimated that only 10 percent or fewer of the species they contain have been described. That means that nearly a million species are thought to inhabit these ecosystems, and this is considered a conservative estimate, with other estimates reaching more than 9 million species. An estimated 600,000 square kilometers (about 230,000 square miles) of coral reefs are scattered over an area of about 150 million square kilometers (nearly 60 million square miles) of tropical ocean. However, not only is this estimate uncertain, but it is also not at all clear how much coral reef area has already been lost as a result of human activities. A loss of at least 10 percent has been estimated, but it is feared to be much greater than that.[22]

Coral reefs grow along continental and island margins within sunlit waters that are beyond the influence of sediments washed from land. Light is required because the living corals harbor unicellular algae, called zooxanthellae, that live symbiotically within their bodies. The algae photosynthesize, producing organic material that is passed into the corals' tissues, providing nutrition, so corals do not have to capture food from the water and their growth is dependent on light. Coral reefs are areas of very high deposition of calcium carbonate, not only from the corals themselves but also from many of the seaweeds and the shellfish associated with the reefs.

The calcium carbonate structure of coral reefs is composed primarily of the skeletons of coral; living corals exist only at the outermost surface. Individual coral polyps secrete calcium carbonate to form chambers with one side open to the water. These chambers are cemented together in colonies that have elaborate shapes characteristic of the particular species of coral doing the building—staghorn coral and brain coral are two examples. Calcareous algae also help in producing coral reefs. Sometimes they cement together the corals; sometimes they form their own calcium carbonate masses over the tops of corals. The base, or dead part, of a reef is a composite of skeletons of the multitudes of coral and calcareous alga species that have been associated with that reef over time. Colonies grow seaward from coasts and upward as sea level rises relative to the reef.

The typical coral reef environment, with its the structural complexity, low level of environmental fluctuations, clear water, and advanced age (modern reefs represent approximately 6,000 years of growth), is conducive to the development of a complex community with a high species richness. Reef communities are highly organized, and species assemblages common to many coral reefs have coevolved. Competition has given rise to a large number of specialized species, which fall into three main groups: the attached organisms—corals, sponges, algae—that give the reef its structure; the fishes; and the cryptofauna, small organisms that bore into, attach to, or hide within the heterogeneous structure of the reef. Most of the diversity and biomass are represented by the cryptofauna, the least obvious part of the community.[23]

The natural species diversity of any particular coral reef is in part determined by geographic gradients. Coral reefs are restricted to tropical waters with diversity greatest near the equator. The longitudinal peak of diversity associated with coral reefs is in the western Pacific and Indian Ocean region, and diversity decreases eastward in the Pacific. The diversity in the Caribbean Sea is not as great as that in the western Pacific but is greater than that in the eastern Pacific, and diversity decreases eastward in the Atlantic Ocean. The diversity in the Indian Ocean is more uniform and relatively high. There are more than 6,000 recorded reefs around the world, and one-third of them are found in the vicinity of Southeast Asia. Of course, the largest reef system is the Great Barrier Reef of Australia. Even though individual coral reefs are isolated from one another, most of the known coral reef species have broad distributions, and endemism seems to be low. The introduction and reintroduction of these species on coral reefs are dependent on other reefs upstream. However, further studies of the cryptofauna might present another picture, since smaller species often do not have reproductive strategies that favor broad dispersal.[24]

Although our knowledge of coral reef diversity comes from relatively recent studies, one coral reef biologist has uncovered information suggesting that the natural species diversity of Caribbean reefs was once quite different from what it is now. This finding seems to confirm that humans have already had profound effects on marine biodiversity and that technology is not always the culprit:

> History shows that Caribbean coastal ecosystems were severely degraded long before ecologists began to study them. Large vertebrates such as the green turtle, hawksbill

> turtle, manatee and extinct Caribbean monk seal were decimated by about 1800 in the central and northern Caribbean, and by 1890 elsewhere. Subsistence over-fishing subsequently decimated reef fish populations. Local fisheries accounted for a small fraction of the fish consumed on Caribbean islands by about the mid nineteenth century when human populations were less than one fifth their numbers today. Herbivores and predators were reduced to very small fishes and sea urchins by the 1950s when intensive scientific investigations began. ... Studying grazing and predation on reefs today is like trying to understand the ecology of the Serengeti by studying the termites and the locusts while ignoring the elephants and the wildebeests.[25]

Coral reefs in other places, such as the Philippines, have also been severely altered.

Nevertheless, studies in less disturbed areas have provided insights into their ecology. Because of the complex physical structure of the coral reef, competitive and predator–prey interactions are intensive and varied. The number and variety of herbivorous fish and invertebrates are important because they graze down the fleshy seaweeds, which would otherwise grow tall enough to shade the coral and calcareous algae, retarding their growth. The exceptionally high biodiversity of reefs, it is postulated, is maintained in part by an intermittent level of natural disturbances rather than by environmental constancy. Natural disturbances on reefs include storms, freshwater floods, sediments, and invasions by schools of predatory fish and groups of invertebrates, all of which may reduce diversity if too severe.

The role of keystone predators is not clear. It was long thought, for instance, that the long-spined black sea urchin in the Caribbean, *Diadema antillarum,* was a keystone species, but when a disease caused a widespread die-off, the expected cascading effects on the diversity of the reef community did not occur. However, it was finally concluded that enough of the urchins had survived that they could still perform the keystone role. This demonstrates both the complexity of the coral reef community structure and the difficulty of identifying indicators of change. Another major predator is the crown of thorns starfish on Pacific reefs, including the Great Barrier Reef. This predator eats coral and can cause dramatic declines in diversity by killing off large areas of the coral on which the community depends. Depending on the extent of damage, recovery may be relatively rapid or excruciatingly slow. There is disagreement as to whether heavy infestations are a natural phenomenon or are induced by human pollution and whether they are increasing.[26]

There is an increase in diversity of coral species with depth, down to a critical depth below which diversity declines sharply because of limited light. One suggested explanation is that rapid growth rates in higher light intensities of near-surface waters cause competition to reach its exclusionary conclusion before other factors can perpetuate coexistence of competing species. There may also be inhibitory factors very close to the surface, such as high temperatures and ultraviolet light.[27]

Researchers have noted that fish species assemblages on coral reefs are not stable; that is, if disturbed, fish species do not necessarily recolonize in the same assemblage. This has led some scientists to conclude that chance colonization may often be the best explanation for the diversity found on a specific reef. Research has also shown that species diversity on coral reefs is increased by intermittent disturbances, usually caused by storms. Others associate nutrients with diversity, suggesting that the greatest diversity is found in the most nutrient-poor waters. Looking at longer time scales, some scientists claim that evolution has occurred more rapidly in warmer waters, resulting in the highest diversity being associated with the warmest surface-water temperatures.[28]

Throughout their very long geologic history, coral reefs have shown great resilience and ability to adapt to natural environmental changes. However, that ability has depended on a rate of change consistent with the rates of growth and natural processes characterizing the reefs. Now, human actions have introduced a rate of change that exceeds the reef's ability to respond. The near-crisis proportions of biological degradation in many of the world's coral reefs are related to human activities, many of which are the result of social crises. The decline in these ecosystems is largely the result of land-based runoff and overexploitation of resources, often by indigenous peoples who are trying to meet immediate food or economic needs and who have few or no alternatives.[29]

Although indigenous peoples have been feeding themselves with fish and shellfish from coral reefs for a very long time, it is only relatively recently that they have begun trying to satisfy national and international markets for food, medicines, and ornamental aquarium specimens. Reefs cannot withstand that kind of pressure. Not only is there too great a demand for reef resources, but also the methods of gathering them are often excessively destructive. For instance, fishing by dynamiting reefs has been favored in areas such as the Philippines, and fish for the home aquarium trade are collected by pouring cyanide into the water to stun the fish, which then float to the top and are captured. More fish die in the process than are delivered alive, and nontarget animals fall victim as well. Furthermore, the traditional systems regulating who fishes where and how much is taken have broken down, causing as much social as environmental distress.

Of equal or greater importance in some reef areas are the consequences of nutrient enrichment and siltation from agricultural and urban runoff, coastal construction, sewage discharge, mining activities, and destruction of rain forests. Corals and their symbionts require clear waters for optimal growth. Supplemental nutrients from runoff promote the growth of free-living planktonic algae, which in turn shade the corals and reduce the photosynthesis of their symbionts. Increased sediment loads washed from land in association with mining and deforestation make the water more turbid (cloudy) and/or smother the corals. Other types of pollution are also important. Chronic oil pollution significantly lowers species diversity in areas where there are refineries, drilling, and tanker traffic.

Tourism, generated by the beauty of the very reefs it threatens, is a source of numerous problems, including pollution, overfishing, damage by boat anchors, and damage by divers collecting souvenirs or walking on the coral. Commercial ship traffic through reef areas is also a serious threat, given that a grounding takes out a significant portion of reef and recovery is slow.

As if all this weren't enough, the effects of global warming threaten many reefs. Corals have a very narrow temperature tolerance range, and many are living at or very close to the upper extreme. Thus, a small increase in temperature caused by global warming might prove lethal, causing vast destruction regionally if not globally. There are already signs of trouble. For several years, widespread coral bleaching has occurred during the warmest seasons in reefs worldwide. Bleaching is a stress phenomenon in which corals expel their algae during periods of physiological duress. A secondary effect of global warming is that the associated sea-level rise could outpace coral growth, leaving some reefs submerged too deep to survive. Finally, there are increasing reports of disease among corals and other reef species. It is not known whether these are related to anthropogenic factors such as pollution or global warming, but it is suspected that there may be a direct connection.

Despite this gloomy portrait, recent studies have given the scientific community hope that at least some of the most diverse reefs, in the western Pacific and Indian Oceans, may not be in as bad a condition as feared. Those that are still flourishing need protection from human activities that have threatened or degraded other reefs, but some scientists believe that reefs have become a priority for the public and policy makers and this may save some of them. The healthy and diverse reefs may then fuel the recovery of others.[30]

Continental Shelf Benthic Ecosystems

Within the coastal zone but deeper than the coastal fringe ecosystems lies a great expanse of gently sloping continental shelf that supports a benthic flora and fauna of unknown and changing diversity. The depth of this environment generally ranges from about 10 meters (33 feet) to about 200 meters (660 feet). There, the gently sloping shelf ends abruptly at the precipice of the continental slope. The widths of continental shelves vary immensely.

Generally, the continental shelf is a soft-bottom environment, although hard rock outcrops may interrupt it. The penetration of light into the water varies with turbidity (cloudiness caused by suspended particles), which in turn varies with the blooming of phytoplankton and with turbulence or outwashing of sediment from shore. Although light may reach the bottom in the shallower and clearer waters of the shelf, the soft substrate does not provide a stable enough environment for attachment. Therefore, except where there are rock outcrops in sunlit waters, seaweeds are not prolific on the continental shelf. Attached seaweeds have been reported as deep as 250 meters (820 feet) in the clear waters of the Bahamas, away from continental margins, but this is the exception rather than the rule. Generally, shelf waters are not as transparent to light because continental margins constitute the zone of densest phytoplankton growth in the ocean and major rivers carry huge loads of sediment that fan out in coastal waters before finally settling and becoming part of the sediments coating the shelf. Therefore, the major primary producers are planktonic microalgae suspended in the pelagic environment. Some animals in the bottom sediments filter food from the water overlying the sediments and feed on detritus and microalgae that sink out of the surface waters. Where the water is shallow and/or where mixing by wind and currents is great, live microalgae may be present throughout the waters, from the surface down to the bottom.

The most noteworthy seaweeds of the continental shelf are the giant kelp forests in waters as deep as 100 meters (330 feet) along the coast of California and in other parts of the world. The roles of grazers and otters have already been discussed relative to keystone predators on rocky shores. Kelp forests are home to a considerable diversity of plants and animals—a variety of very tall and shorter kelp species, invertebrates that live on the kelp and on the rocks, and fish—which are characteristic of these ecosystems and are not generally found elsewhere on the continental shelf. Although the known diversity of kelp forests off California includes hundreds of species, functional and species diversity is a mere vestige of what it was before human exploitation took its toll.

Other diverse hard-bottom communities may be associated with large rock outcrops. The Cordell Bank, off the coast of California, is such an area. It has a high diversity of benthic animals and fishes and for this reason it is protected as a marine sanctuary.

Until recently, scientists thought that tropical areas harbored the greatest total diversity of species associated with shelf sediments, although individual types of fauna might be more or less diverse than their temperate counterparts. Nevertheless, as in many other marine environments, when scientists had a closer look, great numbers of previously unknown species were discovered in several other areas of continental shelf. Now, it is reported that the species diversity of some shelf ecosystems may approach the richness that has been discovered in the deep ocean, though the total shelf area is so much smaller that the balance still tips in favor of the deep sea. Studies in shelf sediments from southeastern Australia and from Norway indicate a great diversity of species, many of them previously unknown and most of them rare. The Australian study found more than 800 species in a rather uniform area only 10 square meters (108 square feet) in size. The distribution of animals in these environments seems well correlated with the grain size of sediments.[31]

Studies of sediment communities in tropical areas have revealed a variety of animals that overlap quite a bit in their apparent roles in the ecosystem and that include many generalist species. This suggests that intermittent disturbances may be important in determining species diversity by preventing competitive interactions from reaching the end point at which weaker competitors are evicted. Hurricanes, cyclones, and El Niño are all suggested as possible agents of disruption, although their effects would depend on the depth of water overlying the sediments.[32]

The continental shelf off the northeastern coast of the United States is one of the most intensely studied regions in the North Atlantic Ocean, yet only about 500–600 species have been described in its sediments. Furthermore, dominance patterns appear to be clearer in this region. These estimates may not include meiofauna; nevertheless the discrepancy could reflect real differences, since diversity may be lower than normal as a result of stress and destruction caused by pollution from several major coastal cities and by commercial trawling for fish. Certainly, it has been demonstrated that both these sources of stress have a marked effect on the benthic community in the New York Bight, the apex of which is New York City.[33]

Epibenthic species, which live on top of the sediments, may be very dense in some areas of the continental shelf. For example, bottom-dwelling fish in shelf areas such as the Grand Banks, off Newfoundland, and Georges Bank, off the northeastern coast of the United States, once supported an unimaginably plentiful fishery of "groundfish," including such species as halibut, cod, flounder, and skate. These species, which represent the transition between benthic and pelagic ecosystems, have a role in both. With the disappearance of most of the large fish from these areas as one after another of the fisheries collapsed, the community has shifted to smaller animals. It is not at all clear whether the large animals will rebound in great enough numbers to dominate once again. There is no accounting that compares species numbers and sizes over time, but clearly there is a substantial difference in the biological character of the ecosystem today.[34]

Vertical currents known as upwelling areas are often associated with low oxygen concentrations in the sediments below. The upwelling waters are rich in nutrients and support a high level of productivity near the surface. The nutrients initiate periods of eutrophication in which high levels of phytoplankton production lead to deposition of unused algae on the bottom and high oxygen consumption as they decompose. Consequently, these sediments may have low populations and diversity of meiofauna and larger benthic species and may be overlain by extensive microbial mats—similar to eutrophic conditions in nutrient-polluted areas inshore.

Shelf ecosystems are subject to a variety of adverse effects from human activities, with those from fishing arguably the greatest. In some of the major fisheries of the world—for example, along the coasts of Alaska, British Columbia, the northeastern United States, and Newfoundland—the continental shelves are trawled for fish and shellfish living on or near the bottom. Bottom trawling destroys the benthic habitat, and repeated trawling prevents recolonization. Continental shelves have also been the site for dumping of all sorts of wastes—industrial wastes, military wastes, sewage sludge, and contaminated dredged materials—and they continue to receive pollutants in the discharge and fallout from industry and other human activities on land. In addition, oil exploration takes place on the continental shelves, accompanied by leaks, spills, blowouts, and discharge of contamination in drilling muds. Natural oil seeps in the sea floor have given rise to a specialized microbial biota, primarily bacteria that thrives on hydrocarbons. These species may prove to be successful invaders in areas where sea-floor drilling leads to oil leaks and spills. Toxic pollutants from a vast variety of human endeavors on land and at sea accumulate in sediments, threatening the diversity of animals that dwell and feed there. Diversity gradients corresponding negatively to pollution gradients are common in shelf sediments.[35]

Human factors may well have become the major forces governing species diversity in continental shelf communities around the world. Disruption of the bottom sediments by occasional small-scale fishing might not have a negative effect if it opened up periodic patches maintaining higher diversity, but the typical repeated disruption and removal of macrofauna (animals of a size visible to the naked eye) over huge areas significantly reduces the diversity—perhaps permanently. The disturbance is too intense, takes place over too large an area, and is repeated in intervals that are too short. Eutrophication at the outflows of major rivers and toxic contamination of sediments are other destructive forces depleting species diversity and exacerbating the problems caused by fisheries. In some areas, pollution from oil exploration may have a similar numbing effect on processes that maintain biodiversity.

Coastal Pelagic or Neritic Ecosystems

The waters lapping exposed coastlines and overlying continental shelves are the most productive marine ecosystems and among the most threatened. These waters, particularly in mid- to high latitudes, are characterized by currents and wave action, upwelling, freshwater outflow from land, and seasonality, all of which serve to keep them rich in nutrients fueling high productivity. Light and nutrients combine to promote the dense growth of a diverse community of phytoplankton (microalgae), which in turn feed grazers—small crustaceans and larvae of numerous types of animals, as well as larger plankton called jellies and salps—and other creatures on up the food chain, from small fish and squid to great fish, dolphins, and whales. Dense populations of predatory fish characterize these productive waters, and marine mammals, seabirds, and humans dip into the water from above, becoming the final link of the chain.

The seasonal aspect of the coastal pelagic environment is critical to maintaining the high levels of production that have fostered runaway fisheries. In winter, the angle at which sunlight strikes the water's surface inhibits its penetration into the depths. With limited light, production of phytoplankton is low, and nutrients build up in the water. Come spring, storms mix more nutrients from the bottom and flooding rivers flush them out along coastlines. Upwelling may also vary seasonally. As the sun moves toward its annual latitudinal zenith, light intensity in the water column increases and the microalgae bloom, fueled by the accumulated nutrients. The zooplankton follow suit, and then the chain of predators.

Some of the higher-level predators are fish that migrate to take advantage of burgeoning food supplies at different times in different places. Even bottom-feeding predators such as cod can move quite long distances, and their movements are fairly predictable. The species diversity of these nutrient-rich coastal pelagic systems is only moderate, but production is very high, which is likely responsible for the lower diversity. Current scientific evidence suggests that there is a humped relationship between productivity and biodiversity: both increase together up to an optimum association, and then, as productivity increases beyond that optimum, biodiversity decreases.[36]

Marine mammals were once important consumers of some of the biomass of fish in coastal pelagic ecosystems. However, they were hunted down to less significant populations—the greatest slaughters occurring in the 1700s and 1800s, avid whaling continuing in the early 1900s, and smaller hunts occurring even today. Their populations are now so small that they consume less than 8 percent of the fish production. Many marine mammals are no longer counted as major parts of the ecosystem—they are treated more as lovable or not-so-lovable tourist attractions, depending on the beholder; when they eat fish, they do so at the risk of wrath from commercial fishers. Even when we love them and protect them from harassment, we do little to protect their food supply. Seabirds are another consumer of the fish in coastal pelagic systems. They tend to eat the smaller fish and thus are not in direct competition with human fishers, but they are victims of entanglement in fishing nets and lines. They also have been hunted throughout human history, some to extinction and many to insignificant populations. Finally, there is the human fisher—the top predator of the coastal pelagic system—an introduced species from land that now has a dominant, global effect on the biodiversity of coastal pelagic ecosystems.

Large marine ecosystems are coastal pelagic ecosystems defined by subtle boundaries. They may be bounded by hydrographic features such as fronts formed by currents and tides that shift at regular or irregular time intervals, they may be upwelling areas, or they may be defined by topographic features of the sea floor such as banks (shoals). Such areas are associated with particular environmental conditions and are characterized by distinct species assemblages. Zooplankton species, for instance, are distributed in communities that can be defined geographically and seasonally in discrete areas of shelf waters. Off the northeastern coast of the United States, separate characteristic zooplankton assemblages in the Gulf of Maine, over Georges Bank, and in the Mid-Atlantic Bight have persisted over several decades. Large marine ecosystems may also define areas of particularly productive fishery species, as, for example, the Grand Banks off Newfoundland, Georges Bank, the Bering Sea, and the North Sea. These are shelf areas where nutrient-rich upwelling waters or currents that stir up nutrient-rich sediments stimulate high productivity.[37]

Tropical shelf waters are generally less nutrient rich than temperate and polar waters, or at least they do not have the seasonal nutrient pulses characteristic of temperate and cold waters. As a result, the phytoplankton are sparser, but they are distributed to greater depths because sunlight penetrates farther into the clear waters. The species diversity of phytoplankton and of some other groups may be greater than in temperate and polar seas, but the populations of the species are smaller. The areas of higher productivity, and consequently the major sources of fish for local human consumption, are the coral reefs with a high diversity of pelagic fish species and areas of river outflow.

Through indiscriminate fishing practices (see Chapter 2), commercial fishers have depleted populations of hundreds of species of target fish and several pelagic shellfish and have had unmeasured effects on untold numbers of bycatch species. The devastating effects of reef fisheries have already been described. Among the most catastrophic northern fishery collapses is that of cod in the North Atlantic and other species of groundfish (epibenthic fish) off the northeastern coast of North America. Cod, hake, and haddock catches decreased by 67 percent between 1970 and 1992. In the early 1940s, California's once lucrative sardine fishery disappeared; by the early 1980s, the anchovy fishery off the coast of Peru had plummeted by 80 percent; and in the 1990s, salmon populations off the northwestern coast of the United States diminished to the point that some runs (distinct populations defined by their spawning rivers) were declared endangered. Halibut and pollock in the Bering Sea and bluefin tuna and swordfish of open seas of the Atlantic Ocean have been fished to various degrees of depletion, not to mention the effects on all the unfished species, such as seabirds, that rely on these species in one way or another.[38]

And yet the pursuit goes on. Now, commercial fishers are moving down the food chain, targeting smaller and smaller fish and invertebrate species such as squid. These would normally be food for the preferred large predatory fish, if the latter were left unfished in an effort to allow them to rebound. But that rebound cannot happen if commercial fishers take their food. Furthermore, the smaller fish include young of the bigger fish, which are not even being allowed to live long enough to produce more of their kind. Some scientists fear that the ecosystem may convert to a new biological community, one dominated by unpalatable species such as jellyfish (or jellies) and very tiny species that humans have no desire or ability to hunt and eat. There are no reports of actual species extinctions caused by fisheries, although the white abalone in California is near extinction, and several populations of some species have been extinguished. There is no question that populations and genetic diversity have been reduced and the roles of species within their ecosystems have changed. In other words, the coastal pelagic ecosystems of the world are severely impoverished. The question is whether they can ever again return to what they once were.[39]

Making the situation even more dire, fish threatened by overfishing are also dealt the double blow of pollution and climate change. Modeling studies of three commercial fish populations off California—northern anchovy, Pacific sardine, and chub mackerel—suggest that mackerel populations may be sensitive to effects of climate change measured in those waters over a twenty-year period and that the other two populations seem to be sensitive to reproductive inhibition by contaminants. In a related but different type of study of fluctuations in herring and sardine populations off the Scandinavian coast, a correlation with climate fluctuations was also found.[40]

Three books published in the mid- to late 1990s give eloquent and poignant testament to the tragedy of the fished seas: Farley Mowat's *Sea of Slaughter,* Carl Safina's *Song for the Blue Ocean,* and Sylvia Earle's *Sea*

Change all carry a grave message about the state of our coastal seas, in particular the devastation of fish populations and ecosystems by commercial fisheries, as well as hope that something might be done to end the massacre.[41] The jacket copy for each of these books likens it in importance to Rachel Carson's *Silent Spring*. Let us hope these books will be more effective. We praise Carson's book, we name buildings after her, we credit her with awakening us to the plight of the environment, and her book is acknowledged as leading to an eventual ban on DDT. Unfortunately, however, action has not been taken on the book's more general message, which was a warning about *all* pesticides that poison the environment. Toxic pesticides have proliferated since then and increasingly threaten birds and other terrestrial and aquatic species; and DDT, even though banned, is still in the environment, wreaking its damage.

Lest all the blame be placed on fishers, some of whom do care and strive to act responsibly, it is important to realize that human activities on land contribute significantly to the problem. In *Our Stolen Future*, which could be considered a sequel to Rachel Carson's book, Theo Colborn, Dianne Dumanoski, and John Myers reveal the full extent and threat of synthetic organic chemicals in the environment today.[42] Pollution is the final blow to deteriorating coastal ecosystems. Synthetic organic chemicals and heavy metals are deadly to many species, especially the eggs and larvae of numerous species of fish and invertebrates. Large areas where there is little or no oxygen—such as a dead zone the size of Rhode Island where the Mississippi River enters the Gulf of Mexico—are barren and will remain so as long as these conditions persist. With continual inputs of fertilizers and poisons from land, shelf systems may have little hope for full recovery if fishing pressure is reduced.

Harmful Algal Blooms in Coastal Waters

Nutrient pollution leads to another serious problem that increasingly threatens coastal ecosystems. The incidence of harmful algal blooms is growing globally. These are blooms of planktonic microalgae unlike the typical spring bloom that fuels the high productivity of coastal ecosystems. They include (1) multispecies blooms so dense that they disrupt the normal food web and lead to oxygen depletion; (2) single-species blooms in which the species is physically unpalatable or nutritionally deficient for the usual cast of grazers; (3) blooms that produce toxins of various intensities that affect animals higher on the food chain; and (4) blooms that are harmful to people. Harmful algal blooms tend to represent a shift from blooms dominated by a complement of diatoms (a type of microalga that only rarely is harmful) historically characterizing coastal marine waters at mid- to high latitudes to blooms dominated by dinoflagellates (also common but more often toxic) and other novel and often harmful types of microalgae. Toxic blooms are sometimes called red tides, named for the color of some common toxic algae. Although some blooms do color the water, not all are dense enough to do so. Harmful algal blooms always indicate decreased diversity of phytoplankton species and, to various degrees, affect the diversity of the rest of the food web. Some toxins affect people who eat seafood that has grazed on the harmful algae, some harm people through direct contact. Toxic algal blooms have been blamed for mass deaths of marine mammals, though it has been speculated that deterioration of the environment had first lowered the mammals' resistance.[43]

The scale and timing of development of harmful algal blooms vary considerably. Some are localized, restricted to portions of bays and estuaries, and others are huge, spread over thousands of square miles of coastal waters. Some occur in the same place year after year, whereas others appear suddenly and are not repeated. Some last only a few weeks, and others go on for years. The causes of the increased incidence of these blooms worldwide are still debated. Some scientists are convinced that it is linked to human-induced changes in coastal environments. Some of the causative organisms may be opportunistic species newly introduced into an ecosystem and able to dominate it swiftly. Others may be around in small numbers all the time, but a certain combination of environmental parameters triggers their rapid growth and dominance.[44]

Besides wreaking havoc with the food web and causing human health problems, harmful algal blooms can have serious effects on fisheries in coastal waters. Several of the more common species that cause red tides produce nerve toxins that do not affect filter-feeding invertebrates (e.g., mussels and other shellfish) that eat them directly but are toxic to vertebrates (fish, birds, and mammals, including humans) that eat the contaminated shellfish. A connection has been established between a toxic microalga and large kills of mackerel off the coast of Argentina. The fish don't eat the algae, but they do eat the phytoplanktonic salps that feed on the algae and, through an efficient food-gathering mechanism, accumulate them in great densities. Besides their causing huge fish kills, there is increasing evidence that some harmful algal blooms have sublethal and chronic effects on fish, including increased susceptibility to disease, depressed feeding, and impaired reproduction. Thus, harmful algal blooms may contribute to the devastation of fish populations.[45]

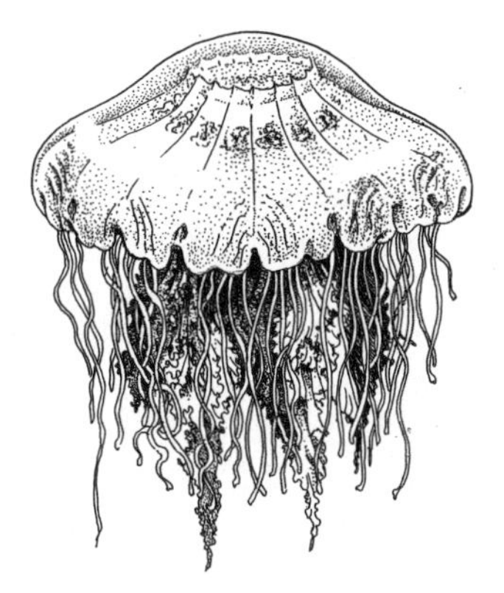

Chapter Five

Open-Ocean Ecosystems

Local- and regional-scale interactions among physical, chemical, biological, and anthropogenic factors dominate the narrow and diverse coastal margins of the ocean. Although interactions on these scales are found as well in the open ocean, its enormous volume and breadth make it critical to global-scale interactions, cycles, and transport involving the atmosphere and the sea. The biota of both the benthic and the pelagic realms are diverse and important and are affected by global processes. The deep-ocean floor houses a very great species diversity, governed by processes on minute as well as large scales. The species diversity of the pelagic ocean may not be as great as that of the sea floor, but its genetic and taxonomic diversity are significant. Both of these realms support living communities that are a far cry from the "desert" the deep sea was once thought to be.

The shelf break, where the continental shelf drops off to become the continental slope, is associated with dominant current and upwelling patterns that differentiate coastal from open-ocean waters and separate their correspondingly distinct living communities; the currents themselves carry an interesting mixture of both oceanic and coastal species that varies throughout the year. Despite these distinctions, the communication between coastal and open-ocean waters is significant.

The western boundary currents along the western edges of the ocean basins (eastern edges of continental shelves), such as the Gulf Stream, are the strongest currents; the greatest upwelling areas and weakest currents are associated with the eastern boundaries, such as the California Current. Major wind-driven currents of the ocean surface are oriented latitudinally, with the winds blowing westward around the equator and eastward at the midlatitudes. The currents thus generated deviate from their westward or eastward orientations in response to the Coriolis force, which is an effect of the earth's rotation causing moving objects or fluids to appear to be deflected to the right in the Northern Hemisphere and to the left in the Southern Hemisphere. When they encounter continents they are further deflected, flowing generally northward or southward along the continental slopes. In the Southern Hemisphere, the eastward-flowing Antarctic Circumpolar Current circles the globe without such deflection, since it flows between the tips of the major midlatitude continents and Antarctica. Elsewhere, east–west and north–south currents join to form large-scale permanent gyres that rotate continuously in one direction (either clockwise or anticlockwise) in the Atlantic and Pacific Oceans. Boundary currents meander, sometimes causing small-scale changes in their patterns and locations throughout the year. These meanders spawn eddies with nutrient-rich, upwelling core waters, and rings, which are waters cut off from one side of the current and carried into the waters on the other side, with temperature being the primary distinguishing characteristic.[1]

Open-ocean systems include two realms—open-ocean pelagic and deep-sea benthic. At the boundary between the benthic and pelagic realms is a layer of water tens of meters thick in which the fauna has both benthic and pelagic characteristics. The horizontal structure of open-ocean waters has already been discussed, but there is also a vertical structure. The upper boundary of the ocean—the microlayer, where the sea meets the sky—is biologically unique. The rest of the open-ocean pelagic environment is composed vertically of the photic (epipelagic), midwater (mesopelagic), and deepwater (bathypelagic) zones, as described in Chapter 3, each with characteristic but overlapping biota. The deep-sea benthic regions include the continental slope and rise, the abyssal plain, seamounts and midocean ridges, and deep-ocean trenches. The biological communities are different in the Atlantic and Pacific Ocean basins, and the polar seas are distinct from these two major basins. The Indian Ocean's biota is more or less a continuum of that in the Pacific basin.

Open-Ocean Pelagic Ecosystems

Open-ocean pelagic ecosystems account for nearly 95 percent of the volume of the ocean, yet less is known about the biology and biodiversity of this entire volume than of any one of the other habitats described in the previous chapter. This huge habitat is compartmentalized, horizontally and vertically, into subhabitats that have discrete environmental characteristics but are separated from one another only by soft physical boundaries—currents or sharp changes in physical properties such as light, temperature, density, or oxygen concentration.

Biotic communities of the open ocean are composed of plankton that drift with the ocean currents and nekton that swim. Phytoplankton (plant plankton) include minute photosynthetic cells, called prochlorophytes; blue-green algae, which are akin to bacteria; and microscopic species of several phyla of true algae. They can live only in the photic zone, where there is enough light for photosynthesis—that is, to a depth of about 200 meters (about 660 feet) or less. Zooplankton (animal plankton) include a great variety of animals, from single-celled protozoa to large invertebrates such as jellyfish that move about on ocean currents. Among the zooplankton, crustaceans (of the phylum Arthropoda) easily predominate; these include numerous species within several categories. There are also representatives of twelve other phyla. Zooplankton include animals that are planktonic all their lives as well as larvae of animals that "grow up to be" nekton or benthos. Planktonic larvae of deep-ocean benthic species are most often found in deep water, but those of some species rise to shallower depths; some are even found in surface waters. Planktonic larvae of both bottom and near-bottom coastal species may be carried considerable distances from shore because of the dynamic exchange between coastal waters and the open ocean.[2]

The life in open-ocean pelagic communities includes mysterious and fantastic forms that are far beyond our daily experience on land or even at the seashore. Among the plankton are gelatinous organisms representing four different phyla, including the true jellyfish, the comb jellies, some squids and other mollusks, and types of chordates not familiar to most of us—ghostly larvaceans and salps, both of which are very distant relatives of humans. The larvaceans have extensive gelatinous envelopes and a tail for slow movement. The salps are solitary or strung out in long, gelatinous chains that are jet-propelled as the individuals pull water in one end and squirt it out the other. Both feed on plankton in the water flowing through their bodies. These are animals clearly adapted to life in a boundless fluid medium, where they never meet a hard surface. Gelatinous plankton come in many sizes and often grow and shrink in response to their food supply. Some grow as large as twenty meters (about sixty-six feet) or more. In the dense, fluid medium of the ocean, where the effects of gravity are minimized, this type of body form is free to develop delicate but expansive and elaborate structures for moving about and feeding.

Other curiosities abound in the open sea. For example, phosphorescence (biochemical light production), which is exceedingly rare on land, is very common in the ocean's interior. We are familiar with lightning bugs on land, but in the sea, phosphorescence is found in myriad pelagic species in many different phyla, from surface-dwelling phytoplankton to some of the gelatinous zooplankton to deep-sea fish that glow with elaborate patterns of light. Fish that occupy the mesopelagic depths (200–1,000 meters, or 656–3,281 feet) characteristically have reflective sides and phosphorescent light organs on their undersides. Phosphorescent plankton are also common in coastal waters.

Filter feeding is another common phenomenon, in fact it is the predominant means of food gathering. Virtually every major taxonomic group of marine animals includes some species that are filter feeders, from small, insectlike crustaceans to great baleen whales. These animals strain their food from the water using a variety of mechanisms, each specifically designed to capture food of a particular size, from microalgae to larger krill. Filter feeding is an ideal adaptation to the relatively sparse distribution of planktonic individuals in the open ocean. This may be why there are fewer fish species here, since most fish depend on hunting their prey one by one. Nevertheless, predatory fish are not absent from these waters. Some have adapted to sparsely distributed food by developing the ability to search out prey with great speed; while some bathypelagic fish have developed phosphorescent baubles and decorations that act as lures, attracting their prey to them.

Although scientists are discovering the answers to many mysteries of deep-ocean life, others remain unsolved and unseen. There seems to be agreement that species are broadly distributed, though many have very small populations wherever they are found and therefore are considered "rare." There has been little assessment of the genetics of populations in different places, so the extent and importance of genetic diversity are not known. With increasing reports of sibling species, genetic studies of open-ocean pelagic species are needed to reveal whether the phenomenon is common in this environment. Furthermore, small animals and microbes appear to be more important than large animals in the open ocean, and speciation and patterns of distribution are poorly known for most of these.

Bacteria and viruses are ubiquitous at all depths, but little is known about their species composition. Prochlorophytes have been found to be living in high concentrations in the waters of the open ocean at a depth of about 100 meters (330 feet), where sunlight barely penetrates. These are the smallest of the photosynthesizers and, along with blue-green algae and microalgae, account for most of the primary production in the ocean and a significant portion of that of the whole earth. Globally, the photosynthesis carried out by phytoplankton is about the same as that of the plants on land, despite the fact that the phytoplankton's total biomass is only 0.2 percent of that of land plants.[3]

The open ocean supports a characteristic pattern of production in all locations, though the species involved and the complexity of the food web vary from one large ecosystem to another. Nearly all life in the sea is fueled by biological processes occurring in the top 200 meters of water or about 5 percent of the ocean's volume. Production begins with photosynthesis, involving the consumption of carbon dioxide in the surface waters, or photic zone. The microalgae and other photosynthesizing microbes are eaten by small zooplankton, ranging from the single-celled protozoa that effectively graze on the smallest phytoplankton to the larger crustaceans such as copepods that filter feed on the larger microalgae. Bacteria digest dead phytoplankton cells. A large portion of the carbon used in photosynthesis is recycled in the surface waters, and more is recycled at the ocean depths. Zooplankton and bacteria return a large portion of the carbon dioxide to the water and pass on the rest as organic carbon to other animals that consume the grazers and bacteria that decompose their fecal debris. This process goes on at successive trophic levels of the food chain, moving into progressively deeper water as the dead organisms and feces fall downward. All along the way, carbon dioxide is returned to the water as a by-product of respiration, and bacteria, which break down the organic material into its elements, return nutrients as well as carbon dioxide to the waters. Finally, some dead and living organisms reach the sea floor, where they are consumed by benthic animals or are deposited as carbon-rich debris. Shells and certain types of cell walls rich in carbonate are resistant to breaking down, and these settle on the sea floor, retaining carbon there for a long time. This movement of carbon from the atmosphere to the sea floor is called the biological pump.

Pelagic species diversity has been studied for individual groups of animals, especially zooplankton, but usually not for an entire community at one time. It is typical for scientists to study one organism category (e.g., fish, zooplankton, phytoplankton) as an indicator of total diversity, which is practically a myth as far as quantitative studies are concerned. To be a good indicator, a functional or taxonomic group should be distributed across the whole ecosystem, and that is why zooplankton are considered good indicators of diversity for the pelagic system. Most assessments of oceanic biodiversity focus on zooplankton or a taxonomic subgroup of the zooplankton community. There are both vertical and horizontal distribution patterns of pelagic species with different processes governing each.[4]

Vertical Distribution

At the very surface of the ocean, where it meets the atmosphere is a layer about 0.5 millimeter (0.02 inch) thick that is distinct from the water beneath in its physical, chemical, and biological characteristics. All marine waters, be they estuaries, coastal waters overlying continental shelves, or the open sea, have a surface skin—the microlayer, which was described briefly in Chapter 2. Because of the properties of water, organic compounds produced by plankton (amino acids, proteins, fatty acids, and others) float or are carried by bubbles to the sea surface and spread out in a thin, concentrated film. Sometimes this layer is evident as a slick on quiet waters, sometimes it can be seen as a long streak or a coating across a great expanse of open ocean, and sometimes it is not obvious at all, but it is generally there. Even if it is temporarily disturbed by severe wave action, it reforms quickly after the waves and winds have calmed. The exchange of gases between the ocean and the atmosphere—an exchange that is of great consequence to our planet's climate—is affected by the microlayer.

The microlayer is also a region rich in life. The organic molecules and nutrients it contains are ideal food sources for a variety of bacteria, fungi, microalgae, and protozoa. These organisms may contribute additional organic substances to the film. Eggs and larvae of animals from deeper habitats commonly occupy the microlayer, as do small zooplankton that feed on the microbial community. It has been reported that fish eggs and larvae can cling tenaciously to the microlayer even when there are large waves. Some of the species that live in this film are found nowhere else and some spend only parts of their lives in this environment.[5]

Photosynthetic pigment (chlorophyll) is often many times more concentrated in the microlayer than in the water below—one study measured it at fourteen times as great. This reflects the microlayer's unusually dense populations of microalgae and photosynthetic bacteria congregating at the surface, where the light energy is the greatest. The microlayer is therefore critical in the consumption of carbon dioxide. Creating a carbon dioxide deficit in the surface layer, which is in contact with the atmosphere, microlayer photosynthesis enhances the ocean's absorption of carbon dioxide from the atmosphere. The microalgae in the microlayer tend to be very small species that are particularly well suited to life in this special environment. Small cells with flagella often dominate, but certain diatoms and blue-green algae are also found in abundance.[6]

Almost the entire oceanic food chain is dependent on photosynthesis, which can occur only at the surface of the ocean, to a depth usually no greater than about 200 meters (about 660 feet). The microalgae and other photosynthesizing microbes are the base of a food chain that extends to the bottom of the sea so that the biology of the surface waters of the ocean ultimately affects the sea floor, even at its greatest depths. A rain of organic material, including living and dead organisms, fecal material, and organic debris coated with bacteria, drops through the water column and is progressively depleted as animals eat it or bacteria decompose it on its way downward. Ultimately, however, enough reaches the sea floor to support a magnificently diverse, though not very dense, community. Species are distributed in layers through the depths, those layers relating largely to physical properties and feeding dynamics.

The stratification of oceanic water into a warmer and lighter mixed layer overlying the colder, denser bulk of seawater is one of the most important features in this regard. The surface layer of mixed water may range from about 40 to 100 meters (about 130 to 330 feet) or deeper, with the depth determined by winds blowing across the ocean's surface. The mixed layer is separated from the rest of the ocean water by a sharp temperature and density gradient, which forms a barrier called the *pycnocline* that resists the exchange of waters between the two compartments. The barrier, however, is not impenetrable to organisms or diffusion of nutrients. Gradients of physical properties are stronger near the surface, so the distribution of species tends to occur on smaller scales than in deeper water. The gradient of light as it diminishes with depth dictates how phytoplankton species are distributed. The scales of distribution of various types of zooplankton tend to be finer in near-surface waters and to increase with depth, and in some cases they may be a function of light intensity because of the distribution of photosynthetic food. Patterns are most pronounced in tropical waters, which are stratified year-round, and less so in polar waters, which are deeply mixed during winter months. Generally speaking, the patterns are characterized by assemblages of species within specific depth ranges. Phytoplankton cells collect to a maximum density just above the pycnocline because this is where, within the mixed layer, the nutrient concentrations are greatest and it is where sinking cells are slowed down or stopped by the change in water density.[7]

Phytoplankton live in the photic zone, with most of the diversity residing in the mixed layer in more or less the top 100 meters. In regions where the mixed layer is relatively stable throughout the year, microalgae may be arranged vertically in species assemblages, with species preferring higher light intensities overlying species preferring lower light intensities. Some species may move up and down within the water as their buoyancy changes in response to physiological changes within their cells. Other species are capable of small-scale motion because they have flagella. This movement through the water undoubtedly enhances their exposure to nutrients. Some dinoflagellates and diatoms (two types of microalgae) are able to migrate upward to maximize exposure to the sun and downward to reach deeper waters of the photic zone that are richer in nutrients. One type does this by propelling itself through the water with flagella, the other by changing its buoyancy.[8]

Zooplankton also occur in vertically separate species assemblages, on scales of 50–100 meters (about 160–330 feet) above a depth of about 600 meters (about 2,000 feet) and on larger scales throughout the rest of the ocean's depths. Species diversity of plankton and micronekton (small fish) tends to increase with depth, with a maximum number of species around 1,000 meters (or about 3,300 feet) for several taxonomic groups. The average size of pelagic zooplankton increases with depth, but the greatest productivity is among the smaller zooplankton in the surface waters. Larger types of plankton, including crustaceans, dominate the upper few hundred meters and give way to micronekton in deeper waters. The average size of fish distinctly decreases with increasing depth, as do the populations and biomass, down to the benthic boundary layer, where the rules change. Bacteria also are distributed throughout the ocean depths and may exhibit vertical zonation, though species are poorly known. Recent genetic identification of species of bacteria in the central Atlantic Ocean revealed a diversity of microbes with broad genetic variation. Bacteria in the mesopelagic zone are thought to break down some sinking fecal material and other large organic particles into finer particles that do not sink readily.[9]

Although patterns of distribution with depth can be discerned, they may be complicated by vertical migrations typical of many species. Daily vertical migrations in response to cycles of darkness and light are undertaken by certain species of zooplankton and fish. Zooplankton may make excursions of tens to hundreds of meters, with most species swimming to the surface during the night and downward by day. Some deepwater fish migrate as well, sometimes coming to the surface at night. The longest recorded vertical migration is by just such a fish—a species in the North Atlantic Ocean that migrates vertically nearly 1,700 meters (close to one mile) every day. Although such migrations are triggered by light fields, their purpose appears to have more to do with the quest for food or avoidance of predation. There is also vertical movement related to different stages of the life cycles of some animals. Certain zooplankton, for example, live near the surface but produce eggs that sink to the sea floor until they hatch as larvae, which then swim to the surface. These kinds of life-cycle excursions are often correlated in some way with water circulation patterns to keep the animal within its preferred habitat during the most productive time of year.

Fish also exhibit distinct patterns of vertical distribution. In the surface waters are large, fast-swimming, predatory fish such as bluefin tuna, swordfish, and sharks. They make long horizontal excursions that may include coastal waters as well as the central ocean, and they feed on a variety of smaller forage fish, squid, and larger plankton. They are sparsely distributed and thus do not account for much biomass relative to the whole ocean, but they are the fish that constitute our high-seas fisheries. In the mesopelagic zone, at depths of 200 to 1,000 meters (about 660–3,300 feet), is found the greatest density of pelagic fish above the benthic boundary. These small fish and larger plankton form a layer that migrates up and down daily. The fish here are characteristically small and shiny, and many have very large eyes, presumably to take advantage of the extremely faint light. They dine on plankton, and many of them undertake daily migrations to feed near the surface at night. The mesopelagic zone is the region of greatest species diversity for fish and some groups of plankton.

The bathypelagic zone, located at depths of 1,000 to 3,500 meters (about 3,300–11,500 feet) is occupied by fewer species of zooplankton and by unique deep-sea fish known for their odd forms and small size. A variety of species are present, most of them characterized by dark color, large jaws, small eyes, and poorly developed musculature; they generally don't move long distances but rely on elaborate structures and phosphorescent lures to attract prey. These species seem to be adapted to take advantage of relatively large prey that is infrequently encountered. Diversity of fish increases with depth, reaching a maximum around 1,000–1,500 meters and then decreasing, although demersal fish (those living on the bottom) are more diverse. Globally, there are an estimated 1,000 bathypelagic (deepwater pelagic) fish species, but the number might be significantly higher. Squid are found throughout the depths of the ocean, and they are an important food supply for high-level predators in pelagic ecosystems. There are hundreds of squid species, most of them relatively small; but one, living in the mesopelagic or bathypelagic zone, grows to a length of twenty meters (more than sixty feet). The "giant squid" is elusive and is known primarily from sea stories, marks on whales, pieces in whale guts, and a very few specimens washed ashore.[10]

Horizontal Distribution

Plankton species diversity at most sites in open-ocean waters is generally as high as or higher than that of most inshore plankton communities. However, the total species diversity of the open-ocean pelagic system as a whole is thought to be very low relative to those of benthic and terrestrial systems. This may be the result of a more homogeneous environment; mechanisms for broad dispersal of plankton; the relative dominance of microorganisms, which generally have low species diversity; limited knowledge about these ecosystems and about the genetics of the organisms; or a combination of all these.

Distinct patterns of diversity distribution with latitude have been confirmed by several studies of plankton that suggest a characteristic pattern that peaks around twenty degrees north or south latitude and declines a bit toward the equator and a great deal toward the poles. Antarctic waters, however, have a higher diversity than Arctic waters. The latitudinal pattern is reportedly found at all depths, not just at the surface, and it applies to both phytoplankton and zooplankton. Sometimes there is an inverse relationship between productivity and diversity, but such a relationship is not always straightforward, and greater species diversity is typically maintained by more constant physical and chemical conditions found at lower latitudes.[11]

There are significant fish populations and species diversity both in the major current systems that delineate pelagic ecosystems and in deepwater habitats. However, numbers of fish species are reportedly much greater in coastal waters, where their populations are also denser—about 50 percent of known fish species are coastal, 12 percent are in the deep sea, and only 1 percent are found in near-surface waters of the open ocean (the rest are freshwater species). Yet there are numerous unknown species of fish in all environments. From what little information is available, open-ocean pelagic communities appear to exhibit relatively high diversity at any given location, but most species are distributed over very large areas, so the total diversity is not considered great. Demersal fish—those associated with the ocean bottom—may be exceptions to this and follow a pattern of uneven distribution leading to increased total diversity.[12]

Species distribution in currents does not necessarily exhibit a pattern because at any one time there may be a number of traveling or expatriate species that have been entrained by the moving waters and are being transported from their native habitat. This is so common that high species diversity is generally found in currents, but sampling programs over long periods of time reveal that a significant portion of such a community is characteristic not of the particular current system but of some habitat "upstream." Some species actually take advantage of current systems to transport themselves from one location to another, with each location favoring a particular stage in the animal's life cycle or its annual feeding and reproductive activities. Entire communities may be inadvertently picked up and moved, with their ambient water, from one place to another in rings and eddies. When this happens, a subset of one ecosystem may end up within some quite different ecosystem, and the visiting species may for a short time become part of the host community. There are also many migrating species that intentionally cross horizontal boundaries between pelagic systems and may even spend part of their lives in estuaries or in freshwater rivers.[13]

The Major Gyres

The horizontal distribution of pelagic communities may occur on a scale of thousands of kilometers. Open-ocean systems are physically and biologically defined by large, stable gyres in the Atlantic and Pacific Oceans: the central gyre of the North Atlantic and the central and subarctic gyres of the North Pacific are the best defined. There is also a subarctic gyre in the North Atlantic, and there are less distinct gyres in the southern Atlantic, Indian, and Pacific Oceans. The biological communities of the Northern Hemisphere gyres are the best studied. The waters in the centers of these gyres contain characteristic biotic communities. Typically, various types of zooplankton, as well as phytoplankton and even bacteria, exhibit patterns of distribution corresponding to large-scale circulation. The gyres circumscribe horizontal areas, but the systems are also vertically stratified, with assemblages of plankton species defined by physical gradients.

The North Atlantic central gyre encompasses a large central area that includes the Sargasso Sea, named for the floating seaweed *Sargassum* associated with its unique community of fish, crustaceans, and mollusks that have evolved fantastic physical traits mimicking those of the golden brown, foliose seaweed. This close-knit assemblage of species of a wide variety of types of organisms is symbolic of looser species assemblages found in many open-ocean habitats. Although they do not mimic one another, groups of species often display distribution patterns that are not independent. Distinct collections of species are repeatedly found together in discrete habitats, especially in particular depth ranges.

The North Atlantic and North Pacific central gyres (also called subtropical gyres) are both anticyclonic gyres in which currents move clockwise in the Northern Hemisphere and counterclockwise in the Southern Hemisphere. This circulation pattern signals an area of downwelling, which is generally associated with low nutrient levels. Thus, the waters in these gyres are generally low in measurable nutrients and relatively high in species diversity. The conditions are relatively constant, and the waters are stratified in such a way that a surface layer more than 100 meters deep persists throughout the year. Nutrients are supplied from deeper waters at a relatively slow and steady rate by processes that are poorly understood. The productivity of phytoplankton appears to be higher than once thought, but grazing seems to keep the standing stock of phytoplankton well trimmed, which may allow for higher diversity of species. The sparse distribution of phytoplankton cells results in relatively clear waters, so sunlight can penetrate deeply, and phytoplankton species distribute themselves along the light intensity gradient. A species-rich community of phytoplankton, protozoa, larger zooplankton, and bacteria in the surface waters accounts for a high rate of recycling that keeps much of the production in that layer of about 40 to 100 meters (about 130–330 feet). However, there is still a significant amount of new production using the influx of nutrients from the air and upwelling, and significant production is lost to the deeper waters through sinking feces and migrating zooplankton. Many scientists once believed that more than 90 percent of the production in these waters was based on recycled nutrients, but later scientific evidence has revealed that periodic upwelling of nutrient-rich, cooler waters from below fuels higher levels of productivity than was originally estimated. This is more consistent with the theory that the highest species diversity is associated with intermediate levels of nutrients and productivity.[14]

The central or subtropical gyre of the North Pacific is the most self–contained, discrete ecosystem in the open ocean. It is generally characterized by sinking waters in the interior, due in part to saltier and therefore denser waters forming at the surface as a result of evaporation. It remains well stratified more or less year-round and temperatures in the surface layer are warm. It is considered the most species-rich pelagic community in the Pacific Ocean. The relatively high diversity of the system is thought to be a result of greater environmental stability than is found in coastal and high-latitude oceanic waters with their seasonal and interannual fluctuation. Grazing and predation also play a role in maintaining diversity.

Two stable assemblages of phytoplankton species have been identified in the surface layer. One group of nearly 250 identified species is associated with a surface zone where nutrients are limiting; the other group, containing only slightly fewer species than the surface community, is associated with a lower zone where nutrients are more plentiful but light is limiting. Both communities are composed of mostly rare species, and only about 20 species in each zone account for 90 percent of the population. There is preliminary evidence that the maintenance of diversity of microalgae within the upper community can be attributed primarily to grazing and that in the lower community to competition among species. Zooplankton exhibit vertical distribution patterns that reflect feeding habits. The surface waters, including both phytoplankton zones, are dominated by grazers; below them are several zones containing assemblages of predators, and near the bottom detritus feeders become dominant. Copepods, the crustaceans with the largest number of species, populations, and total biomass among the zooplankton, are exemplary of this distribution pattern and also include many deeper-water species that migrate to the surface to feed at night. Fish that prey on the zooplankton may play a significant role in determining species composition in the various zones.[15]

The subtropical gyres south of the equator are not as well studied, but they seem also to be characterized by high diversity, though not the same complement of species as found in their northern counterparts. The eastern tropical Pacific Ocean, though not a gyre, is a well-defined area near the equator where nutrient-rich waters converge. It is characterized by upwelling subunits, including the Costa Rica dome, which is among the most productive of tropical waters, and upwellings in the Gulfs of Panama and Tehuantepec. The nutrient-rich Peru Current flows into this system from the southeast. The sources of these rich waters are variable, and they have different temporal scales. Thus, the system is changeable but productive. Its diversity has not been well studied but seems to be only moderately rich, possibly a result of the high productivity.[16]

The subarctic cyclonic gyres in both the Pacific and Atlantic Oceans are characterized by tremendous upwelling of deep water and very strong seasonal cycles in both hydrography and biology. The areas are rich in production but poor in planktonic species diversity. Despite its high productivity, some scientists have suggested that the Pacific subarctic gyre system is less productive than it might be because it is limited by too little of some micronutrient, such as iron, but others believe that biological interactions limit its productivity—that zooplankton graze down the phytoplankton before they can use up the nutrients and bloom to their maximum potential. The mixed layer remains shallow enough throughout the winter to support continued growth of a population of tiny species of phytoplankton; these are grazed by protozoa, which in turn support a significant population of larger zooplankton. As phytoplankton production slowly increases with the increasing light in spring and summer, the zooplankton are able to keep up. Phytoplankton production in the Atlantic subarctic gyre ceases in winter, when mixing nearly reaches the sea floor so that photosynthetic organisms cannot remain close to the surface. When the phytoplankton populations intensify with late-spring stratification of the water, the zooplankton do not catch up easily, and some of the former are lost to the deep ocean. Despite their differences, the two systems support similar production levels and about the same trophic diversity. The main difference is that the Pacific system, where most production remains near the surface, supports more pelagic fish, whereas the Atlantic system, with its sinking phytoplankton, supports more bottomfish.[17]

Currents, Upwellings, Rings, and Eddies

The eastern boundary currents of the world are major upwelling areas and therefore are associated with large fisheries. These are regions where nutrient-rich bottom waters well up to the surface when surface waters are blown westward away from continental margins. These nutrients fuel a highly productive but low-diversity phytoplankton community during periods when the winds relax allowing the development of a distinct surface layer that keeps the phytoplankton in the photic zone. In other words, the high productivity is related to alternation of upwelling and stratification. The phytoplankton support highly productive populations of zooplankton and, in turn, of fish. These are relatively short, uncomplicated food chains. The rich conditions may spread onto the continental shelf, supporting rich seasonal fisheries there, or they may be carried offshore by strong currents during periods of high wind. The species diversity in the currents and upwelling areas is variable and often includes species entrained from deep upwelled water and imported from "upstream" currents. The fish of the eastern boundary currents are dominated by enormous populations and biomass of a few species of relatively small schooling fish. Eastern boundary currents with major upwelling regions are the Benguela and Canary Currents along the southwestern and northwestern coasts of Africa, which support primarily sardines; the Peru Current and upwelling system, which supports a famous anchoveta fishery; the California Current system, which supports anchovy, mackeral, herring, and sardines, which are increasing since their collapse in the 1940s. Data about the California Current have shown a reduction in zooplankton populations associated with a long period, on a scale of decades, of increasing surface water temperatures—an apparent effect of global warming. In this region, during large-scale climate oscillations known as El Niño events, warming of the ocean's surface waters signals a relaxation of wind forcing, causing an interruption of upwelling.[18]

In addition to predictable eastern boundary upwelling regions, there seem to be sporadic upwelling areas along the western ocean boundaries. The best known of these are the periodic summer upwelling off Nova Scotia and the summer upwelling off the western coast of Spain. The Nova Scotian upwelling spreads deep water across the continental shelf and supports what were once the world's richest fisheries on the Grand Bank. Deep water off Spain upwells onto the shelf and into coastal inlets. The enormous natural productivity of this area has been used to support vast rafts of cultured mussels.

Currents and the rings and eddies associated with boundary currents transport numerous species into new areas, where they either expand their distribution or fail in the face of unfavorable conditions. Squid seem actually to ride the currents rather than be swept away by them. On the eastern coast of North America, squid are found in coastal waters from the latitudes of Cape Cod to Cape Canaveral, Florida, but they seem to breed only in the waters off Florida. From there, the larvae enter the Gulf Stream and are carried northward as far as the Grand Banks, off Newfoundland. The adults appear to use northward- and southward-flowing currents and eddies and rings to disperse themselves; when it is time to spawn, they all return south on weaker coastal currents.

Polar Seas

The Arctic Ocean and the seas surrounding Antarctica are special pelagic ecosystems that include a solid phase—ice—in the physical environment. High latitudes, cold temperatures, and ice are common to both ecosystems, but otherwise the two are quite different. To start with, Antarctica is a continent surrounded by deep ocean, whereas the Arctic is frozen ocean surrounded by continents, with broad continental shelves underlying the polar ocean.

The Southern Ocean (around Antarctica) has twice the surface area of the Arctic Ocean, is deeper, and is characterized by more stable conditions. The Antarctic has seasonal pack ice shelves and a great deal of vertical mixing; this mixing is the primary source of nutrients, since there is no terrestrial outflow. The Arctic receives nutrient-laden outflow from continental rivers, and nutrients are mixed up from the continental shelves seasonally. Scientific evidence suggests that the Southern Ocean is poor in iron and that its productivity would be much greater if more iron were made available.[19]

The ice in both polar seas, when present, interferes with light penetration into the waters below. Many primary producers—microalgae, requiring light—have adapted to the situation by living within the ice or adhering to its underside, as do bacteria, protozoa, and some of the zooplankton (e.g., certain copepods and krill). In open areas, microalgae bloom in response to nutrient pulses and are influenced by ocean mixing processes. These organisms form the base of a food chain that ultimately supports numerous invertebrates, fish, seabirds, and sea mammals. Diversity in the ice-covered seas is increased by a system of leads and polynyas—clear areas that arise where ice is driven away by winds or where water rising from the deep creates centers of warmer water, where no ice forms. These are productive areas into which nutrients well up from the bottom, and they are sunlit throughout the light months. Thus, they provide hospitable environments with plentiful food for seabirds and mammals. Many of these open areas appear in the same location year after year, providing reliable places for birds and mammals to congregate. In addition, the heterogeneity of the ice itself may provide for increased diversity of microscopic species living within that environment, though the diversity is not thought to be extraordinarily high.[20]

In the Arctic, the ice also functions as a platform from which seals and polar bears search for food and on which seals breed. The Arctic Ocean's fauna is characterized by mammals, birds, and many different types of fish. In the Southern Ocean, the ice serves the same purpose for penguins and seals, but the fauna of this ocean is dominated by invertebrates such as krill and squid, which support birds and mammals but not many fish. Even greater differences are found in the benthic habitats.

Antarctica, surrounded by the ocean, has pack ice around the edge of the continent, but the ice varies in area by 70 percent over the year. Furthermore, since the ice surrounds a continent, glacial debris consisting of rock and gravel is dumped on the sea floor when the ice melts. This may yield a more heterogeneous sea floor, with a variety of hard surfaces as well as sediments. Antarctica's benthic community has greater species diversity and more endemism than are found in the Arctic. However, its fauna is simpler because there are fewer phyla or types of animals. It does not have crabs, sharks, most benthic fish, or many types of snails, polychaetes (a class of marine worms), large clams, and amphipods (a class of Arthropoda, the phylum that includes insects), all of which are common in the Arctic.[21]

The Arctic Ocean is central, encompassing the North Pole, and surrounded by the northern edges of continents and major islands: North America, Europe, Asia, Greenland, and Iceland. It receives a relatively large inflow of freshwater from the rivers of these continents, with the largest pulse of freshwater introduced during the short period of snowmelt in summer. The ocean is permanently covered by pack ice that freezes and thaws around the edges, changing in size by only 10 percent between winter and summer. The Arctic Ocean's basin is isolated from the rest of the ocean by shallow sills, thus limiting exchange of waters.

Invertebrate diversity off the shores of Antarctica is about twice that of the Arctic Ocean. Diversity in both systems, however, is less than that found at lower latitudes. The cold temperatures do not seem to be a factor in limiting diversity, but they may reduce productivity by inhibiting decomposition and thus slowing the rate of recycling of nutrients.[22]

The Deep Benthic Boundary Layer

The friction created by dense water flowing over the sea floor gives rise to a layer of well-mixed bottom water of a different density from that of the water above. The layer varies in thickness but is generally capped by a density gradient a few tens of meters above the sea floor. It is generally associated with soft bottoms but may also exist over rocky bottoms. It is homogeneous throughout its depth but varies in turbulence and thickness over its area. It is home to an especially rich diversity of benthopelagic species, which are not truly pelagic or benthic but have characteristics of both types of organisms. Some are mainly benthic species that migrate into the water to feed, breed, or escape predation, and some are mainly pelagic species that feed on the bottom. The diversity of animals in this layer is considerably greater than that in the water above, though not as great as in the sediments below. The community is composed primarily of benthopelagic plankton dominated by small crustaceans and larger gelatinous species, including jellyfish and planktonic sea cucumbers. It also includes eggs and larvae of some of the benthic species. Also part of this community are epifauna that roam across the sediments or hard surfaces of the sea floor—animals such as sea stars, brittle stars, crabs, lobsters, sea cucumbers, octopus, and fish. Some of these animals are solitary, whereas others, such as sea cucumbers, shrimp, and schooling fish, are found in clusters.[23]

A variety of bottom-dwelling, or demersal, fish are adapted to skimming over the surface in search of live food or large pieces of organic material, such as dead bodies of large animals, that have dropped from above. Other fish are more truly benthic in behavior, perching or hiding in the sediments in ambush, waiting for prey to swim or drift by. These fish are considerably larger and more robust than the bathypelagic and mesopelagic fish above. They include sharks and rays, eels, and salmonlike and codlike fish. They also include relatives of the bathypelagic angler fish as well as fish found only in this habitat. Benthopelagic fish, which are found on the continental slope and in association with seamounts as well as on the abyssal plain, are not uniformly distributed in depth-defined zones with characteristic species assemblages. They do, however, seem to exhibit the typical deepwater diversity peak at a bottom depth of 800 to 2,000 meters (about 2,600 to 6,600 feet) depending on locality. Although it has been estimated that there are more than 1,000 species of demersal fish, there may in fact be many more, since fast-swimming fish are not likely to be caught in the trawls used for sampling bottom fish. The megamouth shark, for instance, was only recently discovered, though it apparently is present in many locations.[24]

Seamounts are isolated volcanic hills on the bottom of the ocean, which, like reefs, tend to aggregate fish. They create special conditions under which large benthopelagic fish like orange roughy may be found in incredibly high densities, apparently fed by an advection of food by currents related to the surrounding topography. These dense patches of meaty fish have become prime targets for trawl fisheries. However, bathypelagic fish are typically long-lived (the orange roughy, for example, can live for more than 100 years) and are characterized by a slow reproductive process. Reproduction is delayed until late in life, and even then, the reproduction rate is very low. Deepwater trawlers literally mine these dense aggregations of fish, and since they cannot renew their populations to compensate for such rapid cropping, they are susceptible to local depletion over the span of just a few years.[25]

Deep-Sea Benthic Ecosystems

The deep-sea benthic environment appears to harbor the greatest diversity of species in the ocean, likely rivaling the tropical rain forests on land. The ocean floor is shaped by a combination of geologic processes associated with sea-floor spreading and sedimentation of particles, both inorganic (from erosion of rock) and organic (from decomposition of dead organisms). The deep-sea benthic environment begins (or ends) at the continental shelf break at a depth of about 200 meters (about 660 feet). Here, the ocean floor sharply declines as the continental slope, which is composed of sediments overlying the boundary between the continental crust and the oceanic crust. The slope is rather uniform in some regions and quite irregular in others and may be interrupted by deep submarine canyons. Some parts of the slope are prone to slumping, which creates an unstable environment that reduces species numbers. The slope is less than 100 kilometers (62 miles) wide and drops to about 2 kilometers (1.2 miles). At its base, a wedge of sediments slopes much more gently across less than 500 kilometers (about 300 miles) to a depth of about 4 kilometers (2.5 miles). The abyssal plain, which is primarily a soft-bottom environment with organic and inorganic sediments of various grain size, extends gently from 4 to 6 kilometers (2.5 to 3.7 miles) in depth. In most places it is rather featureless, but in some areas it is broken by sharply inclined seamounts, which may occur in chains. Movement of water across the abyssal plain is relatively slow, but periodic benthic "storms" stir up bottom sediments and turbidity currents are associated with huge landslides on the continental slope and in canyons. There are certain finite areas, such as an area on the Nova Scotia continental rise, where such storms regularly interact with swift bottom currents to create a high-energy environment that scours the bottom, leaving a hard surface exposed. The abyssal plain lying between continents is interrupted by midocean ridges, where sea-floor spreading begins with the production of new crust material by volcanic activity and flow from vents along the ridges and associated perpendicular fault zones. In the Atlantic, Indian, and Southern Oceans, the midocean ridge is more or less in the middle of the basin, but in the Pacific, it is considerably offset to the eastern side. This continuous ridge system lies approximately 2.5 kilometers, or 1.5 miles, below the ocean surface. Deep trenches interrupt the abyssal plain in geologically active areas near the margins of continental and oceanic plates. They mark zones of subduction, where the spreading sea floor slides beneath abutting continental masses. Trenches are more common in the Pacific Ocean than in the Atlantic Ocean. All these large features of the ocean bottom have biological implications that are expressed in species diversity.[26]

The animal life on the deep-sea floor consists of (1) large species that move freely on top of the bottom sediments and rocky outcrops; (2) large species that are attached to hard bottom surfaces or embedded in the sediments—the sessile megafauna; (3) small species living in the sediments, often broken into the size categories of macrofauna and meiofauna; and (4) microscopic life, including protozoa and bacteria. Creatures moving over the sea floor are called epifauna or benthopelagic fauna, and these have already been discussed in the context of the benthic boundary layer. Among the sessile animals are sponges; benthic jellyfish, which spend most of their lives in an attached tubelike form and have only a small, insignificant pelagic stage (the reverse of the planktonic jellyfish with which we are most familiar); sea pens; sea fans; deep-sea corals; anemones: sea lilies; deep-sea barnacles; and sea squirts. As fascinating as are the array of animals living on or protruding from the bottom sediments and rocky outcrops, it is the tiny deep-sea animals that contribute most to the sea floor's great diversity. These include unfamiliar variations of numerous familiar phyla, such as a variety of marine worms; bivalve mollusks; gastropods (snails); a selection of arthropod groups; relatives of sea urchins; foraminifera, which are large protozoans with calcareous shells; and several others. There is also a collection of tiny unidentified species coating hard surfaces on the sea floor. There is even a unique group of giant protozoa—single-celled animals transcending size categories from 1 millimeter to 25 centimeters (10 inches) in length.

The food sources for this diversity of life are varied, but all ultimately trace back through food chains to the surface or come directly from the surface. Bodies of dead phytoplankton and zooplankton and fecal pellets rain down relatively uniformly over the sea floor, as do aggregates of organic debris, fecal material, and bacteria. These aggregates are important because they ensure the arrival of a significant amount of the smaller organic particles, which would otherwise disintegrate before reaching the bottom. This rain of small organic material occurs over the entire area of the sea, but quantities vary temporally and spatially with the varying productivity in surface waters. Larger food sources also appear at irregular intervals of time and space. The bodies of large animals such as fish, squid, and marine mammals fall to the sea floor, providing a plentiful though ephemeral source of food for animals that aggregate around them while they last. This temporally and spatially changing food scene provides an environmental heterogeneity that probably promotes higher species diversity. When the populations of whales were much greater, their bodies were likely a very important source of food on the sea floor; they and other mammals still provide food and habitat for a community of opportunistic creatures that move from one to the next as the bodies are consumed.[27]

Most of the diversity of the deep benthic environment is in the form of tiny species of animals that live in the sediments, and its richness has only recently been appreciated. Just how diverse might be the community on the bottom of the sea is a topic of much debate in the scientific literature. Estimates must be based on relatively few samples, with those samples used as clues to extrapolate to the rest of the ocean basin. Credible estimates of species numbers range from a few hundred thousand to as many as a million or a few tens of millions. The disagreement arises from different assumptions used in extrapolating from so few samples to the entire ocean floor. A very few remaining doubters still believe that there are no more species than have already been seen and prefer to believe that sea-bottom diversity is small, amounting to only a few thousand species. This debate is shaping much of the research on the benthic ocean ecosystem.[28]

The first scientists to study the ocean floor envisioned a hostile environment almost devoid of life. They reasoned that the high pressure, cold temperature, and low food supply of this region would allow very few species to survive. Early attempts to sample the deep-sea fauna seemed to confirm these ideas, but unbeknownst to the researchers, most of the specimens collected from the sea floor were lost on the long trip up to the surface. Improved sampling techniques in the late 1960s allowed scientists to recognize for the first time the high species diversity of the animals living in deep-sea sediments in the North Atlantic. Although they found total biomass to be low, both the number of species and the evenness of their distribution were reported to be high—overturning the previous notions that diversity decreased with depth and diversification on the deep-sea floor was limited by environmental homogeneity and sparse, low-grade sources of nutrition. The fact that many of the more abundant species were found to be broadly distributed suggests that barriers between deep-sea communities are less distinct than those between communities in shallow water, but even so the majority of the collected species were rare, and it is still not known how broadly distributed they might be.[29]

Surveys in the Atlantic and Pacific Oceans and the two polar oceans have confirmed that each major ocean basin has its own distinctive sea-floor community of animals and that diversity varies along latitudinal gradients. Many species are ubiquitous within individual basins, but apparently the continents are effective barriers to dispersal and thus the sea-floor biota have evolved separately in each ocean basin. As has been mentioned, there is a depth gradient that appears to be consistent for all ocean basins, although new data suggesting high diversity in some continental shelf sediments may change this concept somewhat. Below the continental shelf, the diversity of larger species increases with depth to a maximum found at an intermediate depth on the continental slope, around 1,500–2,000 meters (about 5,000–6,600 feet). This may be somewhat deeper in the Pacific Ocean. The diversity then decreases with increasing distance seaward on the abyssal plain (which is about 3,500–4,000 meters, or 2.2–2.5 miles, deep) and decreases again in the deep trenches. The smaller fauna appear to reach a maximum diversity somewhat deeper than do larger animals. This may be because the food supply is too small and scarce for large animals in the deepest waters.[30]

Recent samples from the deep-sea floor, at depths of 1,500 to 2,500 meters in the North Atlantic Ocean, suggest a previously unimagined richness of species. From an area of about 21 square meters (226 square feet, the size of a small room) 798 species were identified, representing 171 families and 14 phyla. Of that number, 460 were species new to science. Furthermore, each succeeding sample yielded new species at a steady rate, even after 200 samples had been analyzed. Additional samples brought the species count to 1,597. It is also interesting that the species diversity in samples taken at the same depth varied less than did the diversity of samples taken at different depths, suggesting that some characteristic of the environment that varies with depth, if not depth itself, plays an important role in determining species diversity and may explain the diversity peak at intermediate depths. Extrapolation from these data has led the scientists who did the research to predict millions of species (1–10 million) on the Atlantic Ocean floor, which would give the deep sea an animal diversity approaching that of tropical rain forests.[31]

The North Atlantic is the best-studied region with respect to benthic biota, and the limited sampling that has been done both there and in the North Pacific suggests that species richness is even greater in the Pacific. The distribution of richness there is uneven, and areas relatively close to one another may vary considerably in the number of species they support. This may relate to a number of factors, including differences in the stability of the sediments on the continental slopes, bottom currents, and differences in the productivity of overlying waters. Reduced diversity in the Arctic Ocean is probably due to a combination of the Arctic basin's isolation from the rest of the world's ocean by a shallow sill that prevents biotic exchange of deepwater life, the seasonal productivity of its surface waters, and hydrothermal activity. The Mediterranean Sea, the Sea of Japan, and the Red Sea also are isolated by shallow sills or remoteness from the major ocean basins.[32]

Species richness is probably the result of a combination of temporal and spatial factors on large and small scales. The deep-ocean environment has been stable over long periods of geologic time and is relatively stable, or unfluctuating, on shorter time scales in much of the world's ocean. This stability may have allowed for the evolution of numerous highly specialized species, and it probably accounts for the lower diversity in geologically younger basins—for example, the diversity in the Arctic Ocean is considerably lower than that in the Southern Ocean. Another broad-scale factor that may help account for high species richness is the vast area of the ocean abyss and the ease of dispersal. With relatively few barriers to dispersal, species can repeatedly scatter themselves over a vast area and coexist with numerous other species that are doing the same. This will result in high local diversity. Some ecologists, however, prefer biological explanations, such as the interaction of productivity, competition, and predation.[33]

Most species on the deep sea floor seem to be rare, and their distribution is patchy, which has led to theories based on periodic disturbances and small-scale heterogeneity of the environment. Feeding habits of the deep benthic animals may hold the key to diversity. The disturbance or disequilibrium explanation suggests that diversity is increased by biological disturbances such as predation by large fauna that takes place in a patchy pattern. The dominant lifestyle in the deep sea is described as cropping: the animals feed mainly on detritus raining down from above but may inadvertently consume living particles associated with the detritus, including small benthic animals. The larger, mobile scavengers consume and disperse large particles; whereas small particles such as dead plankton are probably more or less evenly distributed, so the food supply for smaller feeders is fairly consistent. In this setting, the maintenance of high species diversity is attributed to continued biological disturbance rather than to specialization.[34]

The heterogeneity theory suggests that given low productivity, low incidence of large-scale physical disturbance, and large surface area associated with most deep-sea benthic environments, high species diversity is maintained by small-scale biological dynamics. For example, burrowing by bottom-dwelling organisms and the consequent disruption of sediments produces a nonhomogeneous and nonconstant fine-scale topography, and the raining down and settling of organic particulate matter provides an ever-changing supply and distribution of food. Small particles, such as bodies of dead phytoplankton and zooplankton, tend to accumulate in depressions and burrows, and larger objects, such as the remains of fish, whales, and large pieces of debris, cause major changes in the bottom topography and attract congregations of various types of detritus feeders. Thus, a blade of eelgrass falling to the deep-sea floor provides a habitat for a unique fauna. Often these special habitats are ephemeral, so species composition may vary temporally as well as spatially. This biologically induced small-scale variation in the environment has allowed species to specialize and thereby has promoted diversity. For example, species may be specialized in their ways of handling food and in their adaptation to physical microhabitats. Over long periods of time when abyssal environments have been relatively constant, this process of specialization has enabled a great diversity of species to evolve.[35]

Hydrothermal Vent Communities

Unusual deep-sea communities have been discovered fairly recently in areas surrounding hydrothermal vents. These vents have been found in association with all areas of tectonic activity—midocean ridges where new crust is being formed, faults and fractures associated with sea-floor spreading, and subduction zones where the crust of the ocean floor slides under the continental crust on geologic time scales. The vents arise in fissures and fractures that form as molten rock cools when it comes in contact with the water. Cold bottom seawater enters these rifts and circulates under the hardened crust, reacting with the molten rock beneath and emerging as springs of hot water rich in minerals and gases, especially mineral sulfides and hydrogen sulfide, from beneath the earth's crust. These springs may take different forms, including "black smokers," which are very hot and rich in sulfide particles, and rapid gushes of warm water sometimes called "white smokers." A common phenomenon associated with vents is a glow, which may be caused by heat, at the opening where the hot water gushes out. The conditions at these hydrothermal vents can best be described as highly variable, with chemical concentrations and temperature fluctuating over short periods of time and with very sharp temperature gradients, dropping from 300 degrees Celsius or more to ambient temperatures (2 degrees Celsius) within centimeters. A vent has a limited life span, probably on the order of decades or less.

The biological communities of hydrothermal vents constitute a unique food chain. Rather than relying on photosynthesis as the primary process fueling the food chain, these communities depend on chemosynthesis, made possible by compounds, such as hydrogen sulfide, found in the hot fluid around the vents. Highly productive chemosynthetic bacteria form the base of the food chain and support a highly productive community of invertebrate and fish species unique to this environment. Species diversity at vent sites is low—thus far, fewer than 200 species have been identified—but the uniqueness of the species makes them of particular interest. The community is composed of species that are endemic to hydrothermal vents, and many also represent endemic taxonomic families. Despite their low diversity, these communities are impressive to view because of the abundance and density of strange large animals such as giant tube worms, clams, and anemones; unusual stalked barnacles; crabs; shrimp with light-sensitive patches; and small, sinuous fish. The shrimp have attracted a great deal of attention because of their sensitivity to light in such a dark environment. There is speculation that they may detect the glow at vent outlets, an ability that would enable them to find new vents when one dies. However, they may instead be sensitive to the phosphorescence of some deep-sea organisms.[36]

Hydrothermal vents are characterized by low concentrations of oxygen and high concentrations of potentially toxic materials such as sulfides, petroleum-based hydrocarbons, and heavy metals. Most species found in these environments are common to vent communities in general, but a few seem to be endemic to vents of the individual fault lines along which they occur. Because the environment differs so greatly from the rest of the sea floor, vent species are not common to other areas. However, whale skeletons on the sea floor harbor a microbial community that includes chemosynthetic species similar to those found in hydrothermal vents. There is speculation that whale remains may provide refuges or "stepping stones" between hydrothermal vents, possibly accounting for the ubiquitous distribution of many hydrothermal species.

Canyons and Trenches

The benthic fauna of canyons and trenches is less diverse than that on the surrounding abyssal plain and continental slopes. Canyons are unstable areas, subject to slumping sediments on their steep sides. In addition to the unstable sediments, canyons experience rapid currents. These scour the side walls, which tend to be exposed rock, favoring species that attach to hard substrates or hide in crevices to protect themselves from turbidity currents and slides. Canyon waters are generally rich in nutrients, which are carried from the ocean floor in currents that move upward through the canyons. Thus, the pelagic environment in these ecosystems is rich with life. Although their species numbers may be low, these are areas where deep-sea species can be found closer to shore.

Trenches are deeper than 6 kilometers (3.7 miles). The deepest, the Mariana Trench, is greater than 11 kilometers (6.8 miles) deep. These steep-sided ecosystems also have the problem of slumping sediments, but they are not flow-through environments, so water movement may not be as rapid as in canyons. These areas contain their own communities of deep-sea organisms, with a diversity less than and qualitatively different from that of the surrounding sea floor. Reduced diversity is found among species and higher taxonomic levels. Different trenches may have evolved their own endemic species, although they display striking similarity with one another at all higher taxonomic levels.

Human Effects on the Open Ocean and Deep Sea

The pelagic and benthic environments of the open ocean might seem to be out of reach of human influence, but that is not the case. Although this environment is in better condition than the coastal ocean, there is plenty of evidence that fisheries, pollution, and global climate change are increasingly contributing to changes in the ecosystem and that deep-sea mining and exploration for energy sources pose a real threat for the future.

Marine mammals have been hunted relentlessly until they are on the brink of extinction or at the very least play a far less significant role in the ecosystem. Only dolphins have not been targeted by a large-scale hunt, though they have been the incidental victims of hunting for tuna. Large epipelagic fish such as billfish, tuna, and sharks have been fished until they now are threatened and play a reduced role in the ecosystem. High-seas drift nets were widely used in the 1980s, and their use continued until they were banned by the United Nations in early 1992. Miles long and hundreds of feet deep, these ghostly reapers roamed the open seas untended as they "strip-mined" all forms of life too large to pass through the mesh. In a scenario seemingly out of fantasy (for what rational beings would devise implements of such senseless slaughter?), these deadly oceanic gossamers would float slowly through the sea collecting their varied harvests, becoming a veritable sampler of pelagic species diversity. Many of the nets were never recovered; of those that were retrieved, much of the catch was tossed overboard, unwanted and dead.

Now that the epipelagic fish are being depleted, the fishing industry is turning to deep demersal (bottom-dwelling) fish. Dense aggregations of benthopelagic fish, such as orange roughy, have been located in their spawning grounds with acoustic equipment and have been heavily fished because they are such easy targets. Recent advances in mapping of the ocean bottom have resulted in the identification of numerous previously uncharted seamounts. Although these new maps have been applauded as a great contribution to our knowledge of the ocean, they have been viewed by the fishing industry as a pocket guide to potential new fish populations to demolish.

The threat posed by pollution to life on the deep-sea floor has not been given much attention because it has been presumed that at least this environment is out of reach. That, of course, is not true, since the base of the food chain for the deep-sea fauna and much of the food directly consumed by these animals originates in shallow waters. Deep-sea fish in the Atlantic Ocean have been found to be contaminated with persistent pollutants from land.[37]

Because the microlayer accumulates toxic pollutants, the contamination is a serious threat, not only to the diversity of species that are restricted to this layer but also to those animals that depend on it for dispersal of their eggs and larvae. In coastal and oceanic waters alike, this is an important area for the early development of many fish and shellfish, including commercially valuable fish such as cod. Effects of contaminants on eggs and larvae found at the sea surface in sites along coasts include mortality, malformation, and chromosomal abnormalities in fish such as Atlantic mackerel and flounder. There have been documented cases of sharp reductions in successful hatching of eggs and of developmental abnormalities in embryos of Pacific sole and bass collected from the microlayer in polluted waters near urban areas.[38]

Several chemical pollutants have also been detected in remote surface and deep open-ocean locations. Radioactive wastes, which are relatively easy to track in the ocean, provide insight into the magnitude of the problems associated with ocean waste disposal. For example, radioactive materials have been identified in surface waters at Arctic sites where those waters feed the deep ocean, and specific radioactive wastes released from a facility in Great Britain were traced in surface currents all the way to the western coast of Greenland in a matter of a few years. Atmospheric pollutants are now found throughout the ocean. PCBs (polychlorinated biphenyls) and mercury are found in surface ocean waters in concentrations as great as in coastal waters. CFCs (chlorofluorcarbons), which pollute the atmosphere, are actually used to trace the movement of Arctic surface waters into the deep sea and along the ocean floor as they move southward, reaching the Caribbean Sea in only a few years and penetrating the southern Atlantic Ocean over the course of several more years. It may be convenient for scientists to have such a characteristic marker—a synthetic chemical of known surface origin—but the effects on the life in the ocean are not likely to be so positive. There is a real fear that deep-sea species, which have evolved in relatively constant physical and chemical environments, are likely to be especially sensitive to sudden changes in the environment, including the introduction of toxic pollutants.[39]

The effects of global climate change on the open ocean are likely to be related more to changing circulation patterns than to rising temperatures. The revelation that a long-term decline in zooplankton was linked to warming surface waters of the California Current over periods of a decade or more demonstrated how major food web changes may be caused by changing circulation. The warmer waters were a symptom of changes in the circulation that resulted in less upwelling of nutrient-rich bottom water along this eastern boundary current. El Niño patterns in the Pacific give rise to such changes over very broad scales. The effects on marine food chains all along the entire western coast of North America and much of South America are also dramatic. Whether global warming is causing an increase in frequency or intensity of El Niño years has not yet been demonstrated, but it is a genuine concern.

The dumping of wastes in the deep ocean was proposed when ongoing dumping activities proved to be devastating to shallower marine environments. Toxic industrial and military wastes, sewage sludge, and contaminated dredged materials have all been dumped into the sea, usually at shallow-water coastal sites and on the continental shelf. Some have argued that the abyssal plain is an ideal place to receive these wastes and low-level radioactive wastes as well. The argument voiced is that this is a quiescent environment, there isn't much there, and the pollution would not move from its point of deposition and would certainly never reach the surface again. Yet these arguments are invalid, as is evidenced by information in this chapter and numbers of scientific studies of deep-sea circulation and biology. The unspoken reason is that the deep ocean is an area far out of sight and mind and thus the consequences are unlikely to be noticed for a very long time.

Mining of the deep-sea bed for hard minerals such as magnesium has been considered for a long time, and there have been several studies of the economic feasibility of large-scale mining of manganese nodules on the Pacific sea floor and evaluations of the potential effects of such an undertaking. It is not an immediate threat because it is not likely to be profitable in the near future; but should that change, such mining would tear up the sea floor and could destroy large areas of habitat in what we now know is a very diverse marine ecosystem. Our poor understanding of deep ocean biology makes it impossible to predict the long-term effects of mining and dumping.

Exploration of the ocean for energy sources is ongoing and is destined to increase in the future. Oil wells on continental shelf areas are sources of pollution and sometimes of spills and blowouts. As technology permits, deeper deposits can be exploited. And now, another potential energy source has captured the interest of energy developers. Deep-sea sediments contain enormous reservoirs of gas hydrates—icelike solids made up of gas (primarily methane) trapped in a crystal lattice of water. Measurements of such a deposit off the northeastern coast of the United States indicate a reservoir holding more than 300 times the amount of natural gas consumed by the United States in a year, and estimates of global totals suggest twice the carbon found in all the coal, oil, and gas reserves on land. Were technology to become available and economical for the recovery of these hydrates, essentially the whole ocean floor would be available for exploration, but at what biological expense?[40]

Biologists who study the marine environment in general and the deep sea in particular understand that the deep sea floor can be surprisingly rich in species diversity. Nevertheless, many who are not biologists, even scientists who study the nonliving aspects of the deep sea, persist in the belief that the sea floor is a desert. The hydrothermal vents are seen as rich oases in a vast, lifeless sea. More than one scientist has described the deep abyssal plain as a dark, boring, lifeless place, quite suitable for receiving the worst of society's wastes. As recently as 1993, Congress heard testimony to that effect from an oceanographer who had been to the sea floor in a submersible. But in contrast to such sentiments, a marine biologist, Dr. Cindy Lee Van Dover, in her 1996 book *Deep-Ocean Journeys,* which chronicles her experiences as a submersible pilot, confirms the impressions of an increasing majority of marine biologists.[41] Before launching into her fond memoirs of visits to the deep sea where she observed diverse and thriving ecosystems, she acknowledges:

> It is said that the seafloor is a desert, a vast and uniform wasteland, all but devoid of life. Textbooks on the shelf in my laboratory say so. But I know that is not true.

She also echoes an admired predecessor, William Beebe, who in 1934 descended more than 3,000 feet to the sea floor and later wrote *Half Mile Down.* The quote she selects is a suitable rejoinder to those scientists who persist in the belief that the deep ocean is dull and barren:

> If one dives and returns to the surface inarticulate with amazement and with a deep realization of the marvel of what he has seen and where he has been, then he deserves to go again and again. If he is unmoved or disappointed, then there remains for him on earth only a longer or shorter period of waiting for death.

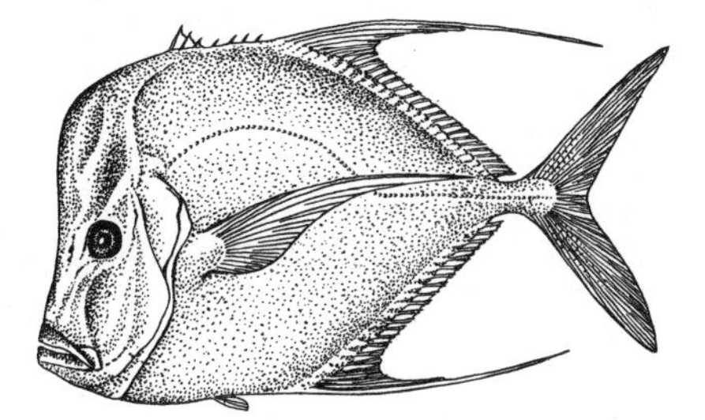

Chapter Six

Conservation of Marine Biodiversity

We rely on science to describe marine ecosystems and biological communities, to determine their functions and sensitivities, and to measure how they respond to human activities. Science fulfills these roles relatively well, with ever-increasing quantity and quality of information and with an improved understanding of what information is missing and where the uncertainties lie. But then we go on to ask science to predict how ecosystems will respond to our activities in the future. The results of those endeavors are less successful and are associated with greater uncertainty and bias. In the face of uncertainty, scientists, governments, and citizens often hesitate to take any action to prevent potential negative effects on ecosystems. The problem is compounded in relation to the ocean, where many changes caused by human activities are missed or ignored. The special difficulties of environmental assessment in the marine environment were noted in 1984 by Rodney Salm and John Clark of the International Union for the Conservation of Nature and Natural Resources (IUCN):

> It is the special burden of marine conservationists that people cannot easily see what happens under water. The sea remains inscrutable, mysterious to most of us. On land we see the effects of our activities and we are constantly reminded of the need for action, but we see only the surface of the sea. Not only are we less aware of our impact on submerged life, but it is also more difficult to investigate.[1]

Although science has learned a great deal about marine ecosystems and marine biodiversity since then, the essence of this statement still rings true, and the more we learn, the more we realize how much there is we do not know.

The first five chapters of this book dealt primarily with scientific information and descriptions of what is known about biodiversity in general and in various marine ecosystems, the processes that maintain it, and the threats to it. The rest of the book addresses what we have and have not done to conserve natural marine biodiversity in the past, the special problems we face because of the nature of the marine environment, and what might be needed in the future to cope with the certain and uncertain effects of human activities. The need to unite science, politics, legal regimes, economics, sociology, education, and ethics in the ultimate protection of biodiversity of marine ecosystems is daunting but clear.

Models for protecting biodiversity have been developed on land, where it is far easier to assess the loss of species. However, because of the differences in the two environments, the effectiveness of transferring particular terrestrial conservation techniques to marine environments is sometimes questionable. For example, individual threatened or endangered species are very often the focus of conservation efforts, but that approach is of limited value, particularly in the sea, where it is useful only for those few species we know to be in trouble and are able to monitor. Terrestrial species are more familiar and their distributions and status are more easily determined, whereas the status of most marine species is unknown. Furthermore, there is little wisdom in waiting until a species has been reduced to ecological insignificance to start protecting it. Methods for protecting ocean biodiversity from the negative effects of human activities should be developed in the context of the unique character of the marine ecosystem and our relationship to it.

Properties of the Marine Environment that Affect Conservation

Some of the qualities that distinguish marine ecosystems have policy implications for the protection of biodiversity. Among them are the following:

- *Water is the universal solvent.* The fact that water can dissolve so many materials means that most pollutants in ocean waters are in a reactive state and readily interact with the biochemistry of organisms there. Pollutants may have effects at low concentrations. Furthermore, the measurement of concentrations in the water does not reflect the total exposure of marine organisms to contaminants, which flux in and out of solution, are continually taken up by the biota, and move around with changing patterns of diffusion and water circulation. To protect marine biodiversity from the effects of pollutants requires strict regulation or prevention of emissions at the source. In all but the shallowest sediments, retrieval of pollutants is not an option as it may be in terrestrial ecosystems.

- *Dissolved pollutants tend to concentrate in the sea surface microlayer and in ocean sediments.* As potential sources of contamination into the food web and into the rest of the marine environment with consequences for biodiversity, these areas, rather than the water column between them, should be the primary focus for contamination standards.
- *The ocean is a fluid with complex circulation patterns.* The result of the movement of the ocean's water is that dissolved and suspended pollutants can be carried long distances from their sources. Because they are generally reactive in water, these substances may affect organisms all along their path of transport. Water quality standards are therefore less protective of the biota in the open ocean than in closed bodies of water where dispersal is more restricted. This also implies that regulations of concentrations in ocean water are of marginal value and regulations should instead focus directly on the sources of pollution. International efforts are often necessary for effective prevention of pollution.
- *Planktonic lifestyles and planktonic larvae of nonplanktonic animals are common in the ocean.* Planktonic organisms drift with the currents and therefore do not remain in one location or even in one ecosystem. The implication is that protected areas should be very large and defined by circulation patterns, not political boundaries, and they should be associated with protection of important source areas of larvae for replenishment of communities. At best, the protection will still be incomplete.
- *Species are often difficult to distinguish.* The increasing frequency with which sibling species have been found imply that we are a long way from being able to assess the status of many species with any certainty. A discovery reported in the mid-1990s demonstrates the effect uncertainties and errors in species identification can have on marine resource management. The common coastal mussel, *Mytilus edulis,* which has been monitored for many years as an indicator of changes in pollution levels in coastal areas, was discovered to be three species. This could mean that physiological differences between species account for some of the "changes" in bioaccumulation of toxic chemicals measured in the monitoring studies. It also suggests that it may be difficult to identify rarity and endangered status in species if groups of sibling species are not being accurately distinguished.[2]
- *Populations may be as important a unit of diversity as species in the ocean.* This suggests that protecting only a single population of a threatened or endangered species often will not adequately protect the genetic variability necessary for the long-term survival of the species.

It also means that distinct endangered populations should be protected even if the species as a whole is not considered endangered. We are seeing the beginnings of such strategies in, for example, the protection of separate runs of Pacific salmon.

- *Food webs are more complex in the ocean than on land.* This is due to the significant added dimension of depth, to the fact that many marine animals spend different stages of their life cycles functioning at different trophic levels, and to the greater variety of feeding techniques in marine ecosystems. Such complexity makes it more difficult to predict the consequences of various human activities on marine community diversity. Removal or reduction in numbers of a particular species affects all the species that eat it or are eaten by it at all the different stages of its life cycle, as well as those that might consume it after its death, and the effects cascade downward. It is therefore difficult to predict all the possible effects of a particular activity in the marine environment. Such uncertainties require cautious policies.
- *Many marine species have a high reproductive output.* Because prolific reproduction patterns lead to great fluctuations in populations, it is difficult to separate anthropogenic from natural causes of population declines. The implication is that regulations regarding such species should be conservatively based on estimated effects given a minimum natural reproductive success.
- *Many marine species have short life cycles.* This also gives rise to large population fluctuations and causes populations to respond quickly to environmental changes. Besides making it more difficult to predict responses to human activities, the short life spans of certain species can result in unexpected events, as in the case of intense, sometimes harmful, algal blooms.
- *The ocean ecosystem is characterized by communities that shift in space and time.* Migrating species are only a small part of this picture. Shifting and fluctuating currents and patterns of turbulence often redistribute pelagic communities seasonally and result in changes in local populations and communities over time. Communities and populations of benthic organisms also rely on recruitment of larvae from other areas. This makes it difficult to effectively assess and manage the taking of living marine resources.
- *Processes that affect biodiversity in the ocean operate on vastly different spatial scales.* Effects may be felt in areas ranging from a few meters to thousands of kilometers. The movements of organisms include short movements related to interactions with other organisms, long migrations related to seasonal feeding or reproductive behavior, and long-distance transport with water currents. For example, sea turtles live as adults in the open ocean but lay eggs on the seashore and eels live as adults in streams and estuaries and migrate to the ocean to spawn so both species are affected by environmental conditions in

ecosystems great distances apart. Similarly, pollutants may be carried over long distances, as in the case of persistent organic pollutants that originate in the tropics and end up in the Arctic; they may be restricted to more local areas, as in the case of nutrients; or they may be sequestered in sediments close to their point of origin but released slowly over long periods. Local human activities may have far reaching consequences over distance and time.

- *It is difficult to study and monitor marine ecosystems.* This means that it usually is difficult to assess the status of species and communities and to measure the effects of human activities and their regulation. The effects may not be seen until the integrity of an ecosystem is compromised or the system has failed and recovery is questionable. Sometimes there are early warning signs, but it is easy to ignore them. The selection of indicators to monitor is of the utmost importance. Moreover, requiring difficult inventories and status assessments before establishing protective measures may delay those measures until it is too late for them to be effective. This dilemma gives rise to heated debate between users and protectors of marine biodiversity.

Approaches to Protection of Marine Biodiversity

Given the foregoing considerations it is difficult to design effective marine conservation plans. If we cannot predict, how can we protect? In the end, our only option is to regulate human activities with sensible precaution based on minimizing human-induced environmental change and, more importantly for each of us to take personal responsibility for our own activities and our own roles in the global ecosystem.

Until such an ethic prevails, however, there are a number of approaches to managing the environment that have met with limited success. Employing tools such as laws and regulations, management techniques, and international agreements, these approaches are used to curtail human activities that have harmed the environment or to try to correct damage already done. They are generally used to address existing problems, but they could be applied in a more anticipatory way to achieve greater success.

These approaches include the following: identification and protection of rare and endangered species; designation of protected areas, where human activities are restricted to protect the environment; integrated coastal zone management, in which a complex of interactive rules governs the multiple uses of coastal areas, including both land and water; regulation of fisheries; regulation and prevention of pollution; environmental monitoring and research to assess ecosystems and human impacts upon them; risk assessment to determine whether regulations are needed; a variety of economic incentives to reduce activities having negative environmental effects; and the repair of ecosystems that have been damaged.

Protection of Species

One of the most common approaches to biodiversity protection is through the identification and differential protection of endangered, threatened, and rare species. This approach was developed on land, where it has saved a number of species from extinction—at least among the more conspicuous species. In the sea, it has been used with similar success to protect a number of marine mammal species from extinction, and several other marine species, including certain fish and corals have been listed for protection. Most recently, populations of salmon along the Pacific coast of the United States have been recognized as endangered, and means for their protection are being developed.

Several species of whales have been brought back from the brink of extinction by international protection, which ended whaling by the vast majority of nations and in effect restricted it for those nations that refuse to stop (Norway, Japan, and Iceland). Nevertheless, almost none of these species have recovered to prehunting populations. For some species, such as the right whale, protection has come too late, and extinction of some genetically distinct populations seems inevitable. Although fleets still known to be hunting claim to take no endangered whales, a new technique for genetic analysis of whale meat has revealed the presence of endangered species in markets in Japan and elsewhere.

There are serious problems with protecting marine biodiversity by protecting endangered species. One is that the majority of endangered species will not be identified or protected. Most species remain unknown to science and therefore are of unknown status; and even for known species, it is very difficult to determine the distribution and status of populations. Consequently, many endangered species will go unlisted and unprotected, and so may their habitats. Furthermore, enforcement of protection on the high seas is difficult at best. It is easy to escape detection when illegally killing protected marine species, and since many protected species are caught accidentally by commercial fishery operations, there is usually a limit rather than a prohibition on such "incidental" kills. Enforcement of those limits is next to impossible.

Another problem with the endangered species approach has to do with the roles of species in their ecosystems. Once they are endangered, the species have ceased to play a significant role in the functioning of their ecosystem. Whether protecting them at that late stage will enable them to resume a key position is often questionable. Furthermore, once a species reaches the stage of endangerment, other species that interact with it will have been affected by its diminishment, and the structure of the community may have permanently changed. Instead of being protective of the ecosystem, the endangered species approach is a last-gasp effort to hang on to a repository of genetic information that will otherwise be lost forever. It may be essential in cases that have already progressed to that stage, but it is not appropriate as a front line of defense in either marine or terrestrial ecosystems.

Following the same line of reasoning, living gene banks, such as zoos and aquariums, are important only as a final, desperate effort to save important genetic material and, perhaps, to make us feel less guilty. Zoos have in fact been successful in preserving some severely endangered species. Public aquariums, on the other hand, have not yet served as effectively as gene banks, particularly for marine mammals. Both aquariums and zoos have been successful in educating members of the general public about many species they will never see in the wild, and it is hoped that as a result, people will want to protect them in their natural habitats. A recent poll indicated that most Americans believe that one of the most reliable sources of information about the marine environment is their local aquarium.[3]

Protected Areas

Another approach to the conservation of biodiversity is establishment of protected areas. These may be marine sanctuaries meant to protect biodiversity while serving other nonconflicting uses, or they may be areas set up to restrict or prohibit a particular activity, such as fishing or ship traffic, because the ecosystem is deemed to be particularly sensitive to that activity. Large marine sanctuaries may be established to protect big sections of marine habitat, such as coral reefs and other areas noted for their distinctive and abundant biodiversity, but regulation and management schemes differ among them. Regulations depend on what are seen as the major threats to biodiversity and what the public will support or tolerate, two criteria that may be in conflict with each other.

Among the most effective protected areas are those associated with coral reefs because these ecosystems are more like submerged lands and their biological communities tend to be relatively stationary, rarely leaving the home reef system. Nevertheless, in an area that contains many reefs, populations of fish and benthic reef organisms replenish one another in an upstream-to-downstream pattern. The more reefs that are protected in a region, the better, and protecting up-current reefs will benefit unprotected reefs down-current as well.

In multiple-use areas, exploitation of resources may be allowed within sanctuary boundaries; the challenge is to make sure the exploitation is restricted to sustainable levels, a topic discussed in the section on fishery regulations. Even in those rare areas with total, parklike protection, the effects of visitors must be minimized if protection is to work. Multiple-use marine protected areas are often seen as a vehicle for establishing effective integrated and adaptive management programs. Although these are useful models, it is time to advance beyond the model stage and increase integrated management to entire coastlines.

Large areas are usually preferable for multiple-use reserves, but the large size makes enforcement difficult. Patrolling a large marine area is expensive and must be done by powerboat, an activity that itself can interfere with the purposes of the sanctuary. Education, therefore, is an important part of a protected area program because if it is to be effective, people, especially potential encroachers, must be convinced that there is good reason to obey the restrictions.[4]

No-fishing areas, which are still astoundingly rare in the marine environment, are set aside to protect important spawning grounds or to provide a refuge area for species that are heavily fished in surrounding waters. Here, they have an opportunity to replenish their numbers and grow to a larger size, which benefits the fishery outside as well as the fish. For instance, it has been proposed that the Channel Islands National Marine Sanctuary establish ecologically independent fish refugia surrounded by areas that can be fished, with the aim of enhancing fished populations and protecting community structure.[5]

Despite the popularity of the protected area approach in the national and international conservation community, these areas are not always defined and protected in a way that maximizes benefits to the ecosystem. For instance, protected areas are often defined first by shoreline features or boundaries, then by bottom topography, and rarely by water circulation patterns. Yet water circulation is probably the feature most important to pelagic communities and is critical to benthic communities as well.

Even if the area is well defined by circulation and has well-designed protective regulations, it is still subject to pollution carried from afar on currents and may fail to recruit larvae if unprotected source areas and immigration paths are damaged. In fact, larval dispersal is severely affected by fragmentation of the coastal environment by human activities that alter its chemical or biological character. In addition, biota that leave a protected area as migratory adults or as planktonic larvae lose their protection. It is much more difficult to predict the paths and destinations of drifting and migrating marine animals than it is for migrating terrestrial animals. Although establishment of protected areas should be strongly encouraged within the larger context of managed coastlines and broad fishing and pollution regulations, it would be a mistake to think that marine biodiversity can be saved solely or even largely by isolated protected areas.

Integrated Coastal Zone Management

Several scientists have made the observation that the world is heading toward a future of total environmental management by humans. Accordingly, in a process called command-and-control management, first the landscapes and then the seascapes will be manipulated so that humans determine which species grow where and for what purpose and, very likely, which species will go extinct and which will be preserved—ecosystems will be dominated by horticulture and agriculture, even in the sea. Although this is an extreme and ill-advised approach, some level of management of coastal areas is inevitable, whether to protect the coastal environment or to direct the ravage and plunder of its natural resources and habitats.[6]

The coastal areas of the world are under intense pressure from burgeoning human populations. Rapid urbanization, residential and recreational development, and many inland activities have a profound effect on coastal ecosystems. It is clear that piecemeal regulations and management practices will not be effective. For example, extensive inland agriculture contaminates groundwater and surface water, and rivers carry this contamination into estuaries and from there into open coastal waters. Ocean and air currents carry some contaminants much farther into the open ocean, to other coastal areas, and into the deep sea. Industrial emissions, sewage treatment discharges, coastal construction, and a host of other activities have synergistic and accumulative effects on the coastal environment. A comprehensive management plan that addresses them all is likely to be more effective than regulation of each in isolation.

Integrated coastal zone management is still conceptual, but there are attempts to make it a reality in some areas. In the United States, for instance, the state of Oregon has developed and enacted a comprehensive coastal management plan for protection of the coastal environment. California is a few steps behind, but has developed a plan and hopes to enact it. Parts of the coasts of California and Washington State are protected by marine sanctuaries. If the three coastal states could adopt a joint plan, an idea that has been proposed by Oregon officials, integrated coastal zone management could become a reality on the West Coast. In other parts of the world, countries bordering enclosed or partly enclosed seas have developed cooperative plans to protect the marine ecosystems they have in common. These agreements are developed very slowly and with difficulty, but sometimes the dire state of a region's marine environment prompts strong enough measures to have a positive effect. However, it often remains easier to agree on regulation of individual categories, like fisheries or certain types of pollution, than on integrated regulation of the whole.[7]

Cooperative management programs for offshore marine systems are also being developed on regional scales. Known as large marine ecosystems (LMEs), these units, which are often defined by ocean current systems, are biologically meaningful. Generally, it is the extraction of resources that is managed, not land-based activities that affect the system. Such areas include the Bering Sea, the northeastern Atlantic Ocean, and banks off the eastern coast of North America.

One management alternative that has been rarely used but is likely to become more common in the future is zoning. Multiple-use protected areas wrongly give the impression that regulation of the marine environment should be reserved for areas that need special protection. Instead, the goal of integrated coastal zone management should be the zoning of a continuous coastal area that incorporates both land and sea. Such zoning, if done responsibly, would include some areas where almost no activities are allowed, but it would also contain a system of zones where different levels and kinds of activities are permitted, with some areas having very few restrictions. The Great Barrier Reef Marine Park, a very large protected area stretching along nearly the entire length of the Great Barrier Reef, off the northeastern coast of Australia, is managed in this way.

Fishery Regulations

Commercial fisheries—one of the most serious threats, if not the major threat, to marine biodiversity in coastal waters and to some species in open-ocean waters—are largely unregulated. Until recently, the only restrictions were related to national boundaries and territorial waters. Most nations do not allow foreign fishing fleets within their territorial waters except by formal agreement, and they often blame foreign fleets outside their waters for depleting fish populations that migrate across or straddle territorial boundaries. Acknowledgment of the need for strict regulations has led to the negotiation of treaties that govern fishery resources in international waters. Individual countries have generally done a very poor job of regulating fisheries in their own territorial waters.

Although overcapitalization (too many boats with too much equipment) and advanced fishing technologies are often cited as playing a large role in overfishing, there has been little attempt to regulate them. In areas where fisheries have collapsed, such as New England, there are some government programs in place to buy out commercial fishers and take their boats off the sea. This counters previous subsidies and programs that enabled them to buy and equip the boats in the first place. Nevertheless, in areas around the world, huge industrial trawlers and other factory fishing vessels continue to overfish, and are poorly regulated, driving more fisheries to collapse and marine ecosystems to failure. In other fisheries, giant drift nets in the open sea were so egregious that the United Nations community finally could ignore the situation no longer and passed a resolution to ban them. However, smaller drift nets and gill nets are still used in coastal waters, and no ban can rid the ocean of the abandoned or lost "ghost" nets, large and small, that continue to roam the sea, reaping their grim and wasteful booty.

Other technologies leading to overkill include aerial spotters and acoustic fish-finders. Even satellite data are now available, processed and distributed by private companies that sell the real-time information to fishers so that they can find and fish down the remaining large assemblages of fish that live near the surface. New maps of the sea floor identify numerous previously uncharted seamounts, so now the congregations of slow-growing, slow-reproducing fish that collect in these deepwater environments—some of them gathered there to spawn—can be easily located and rapidly scooped out until they are depleted. The assumption is that all the living resources of the sea are free for the taking. The only challenge is to find them, and that is becoming easier and easier.

There is an urgent need for international regulation of gear, tracking equipment, and number and size of boats. In some places, such regulations would also have positive sociological effects. Rural artisanal fishers in small boats, who do not cause as much damage to ecosystems, are being driven out by commercial fishers in big boats with high-tech equipment. There is a real need to reverse the trend not only to save the marine ecosystem and fish populations but also to put the catch in the hands of those who need it most for personal and community consumption.

Classic types of fishery regulations, including catch limits, seasons, and a system of open and closed areas, should be employed more than they are now. Even these have to be carefully thought out. One example of how regulation of a fishery can be as senseless as no regulation at all occurred in the halibut fishery off the coast of Alaska. To prevent depletion of the stock, regulators imposed a one-day season during which any number of boats could fish and there was no limit to the number of fish they could catch. The result was twenty-four hours of insanity each year as too many boats competed to find the spots where fish were aggregated; once the lucky boats were in place, their crews would work nonstop until the end of the day-season. Once this dangerous activity came to an end, all the fish had to be processed and frozen. Fresh halibut became a rare commodity indeed, available for only about a week or two. In the face of such irresponsible management, new approaches are being tested.

One regulatory option that is favored internationally is the use of ITQs (individual transferable quotas), whereby the owner of a fishing boat purchases rights to fish a particular plot of ocean, from which other boats are prohibited. Proponents of this method claim it reduces fishing pressure and makes fishing safer. Opponents object, however, that such "ownership" of a parcel of the ocean breeds privatization of a public resource that is valuable for many purposes other than fishing and that has intrinsic value as a naturally functioning ecosystem.

Preferable regulatory policies include a combination of direct limits on catch and gear with a system of closed areas. For example, long-lived and slow-reproducing fish can sustain no more than a 10 percent culling of their populations. Even then, methods such as bottom trawling can destroy habitat, reducing populations even further. Besides, a simple catch limit does not guard against the targeting of large spawning fish, which has a greater negative effect on sustainability of populations. Consequently, it is also important to have well-selected zones that are completely off limits to fishing—marine reserve areas that will protect a wide variety of species, including those in danger of being rapidly overfished. In addition to restrictions on fisheries, there should be an end to subsidies that encourage and equip them.[8]

Establishing such regulations has proven difficult at the national level and will therefore be even more difficult internationally. Marine fisheries have a long history of nonregulation, based on a "take all you can get, it's free" mentality. National governments and international organizations such as the Food and Agriculture Organization of the United Nations (FAO), which monitors global fisheries, are struggling to decide what to do about the world fishery crisis. In large part, they are concerned that overfishing will lead to a severely reduced food supply, but they are also concerned, at least in principle, about effects on ecosystems. There is a recognized need for an international structure to develop a long-term framework for attaining the goals of biodiversity protection in the context of sustainable fisheries and short-term management measures to implement those goals.[9]

National and international regulation of fisheries rarely addresses nontarget species of fish, mammals, birds, turtles, and invertebrates that are killed as bycatch. Furthermore, the more complicated problem of identifying and protecting species whose populations may be depleted because of their dependence on fisheries species is not addressed by typical fishery management programs. Establishment of an international framework might provide a vehicle for addressing both of these serious problems.

Pollution Regulation and Prevention

Pollutants in the marine environment fall into two categories for the purposes of regulation: nutrients and toxic substances (known in the United States as "toxics"). Excess nutrients lead to eutrophication and reduction of biodiversity, as described in Chapter 2. The regulation is in theory fairly straightforward by placing limits or water quality standards on concentrations of nutrients in the aquatic environment. It should be easy to determine whether standards are being met in a particular body of marine water, but if nutrients are rapidly taken up by blooms of algae, concentrations in the water may not reflect the true magnitude of inputs. Furthermore, controlling the inputs may be difficult. Runoff of nutrients from agricultural fields and forestry areas is best reduced by reducing the application of fertilizer, by planting or leaving wooded buffer zones between fields or logging areas and streams, and by limiting the access of domestic animals to streams. Discharges of nutrients from sewage treatment plants can be reduced by more advanced treatment (tertiary treatment removes nutrients but is very expensive) or by innovations such as constructing wetlands to filter sewage water or using modern composting toilets instead of water-based sewage collection. Some nutrient-rich compounds in emissions from fossil fuel combustion can be trapped and removed by advanced technologies, or the amount of fuel burned can be reduced. Some solutions are expensive and others are not, but this problem is manageable with a great deal of diligence—something that has not yet been applied.

The problem of toxic substances, however, will be addressed only with great difficulty and strong commitment. Toxic metals are natural substances that become a problem when they are amassed in high concentrations. This happens regularly with the extraction of metals from the earth, an activity that is not likely to stop. Oil, which gets into the marine environment through numerous careless actions, presents the same challenge. Synthetic organic pollutants are more insidious because they are numerous and difficult to monitor and in many cases have biological effects at very small doses. Many persist in the environment for long periods of time, causing bioaccumulation and biomagnification in food chains.

Since pollution is a major threat to marine ecosystems, the policies of human societies and governments with respect to water pollution are critical to the conservation of marine biodiversity. Policies regarding the introduction of contaminants into the marine environment reflect three general approaches. The first approach, intentional use of the oceans for disposal of wastes from land via pipelines or dumping from designated waste disposal vessels, was once routine but is now progressively being curtailed or banned, although it is still favored in some countries. The second approach, regulation of discharges to maintain "acceptable" levels of contamination that are not expected to cause significant harm to the marine environment, is favored by most governments, though it is flawed in practice. The third approach, prevention of pollution, has been adopted as a goal internationally, but little has been done to implement it.[10]

The determination of acceptable levels or standards for toxic chemicals is always controversial. For some, it is an attempt to scientifically define just how much contamination an ocean system can withstand before it is seriously damaged. This limit, referred to as the "assimilative capacity," is based on the belief that contaminants are not harmful at low levels and there is a threshold of concentration for each chemical below which no harm is done. Typically, determination of these thresholds is based on laboratory tests, known as bioassays, that measure the numbers of deaths of test species in response to different concentrations of the chemical or sometimes a mixture of chemicals. This approach is rarely able to incorporate the additive and synergistic affects of a variety of pollutants from a variety of sources.[11]

Yet in marine environments, chronic nonlethal or delayed lethal effects are widespread and are responsible for biological impoverishment of polluted ecosystems. These effects are caused by long-term exposure to low concentrations of chemicals and/or interactions among them. Recent research has revealed numerous low-level effects of many synthetic organic pollutants. Some, for instance, mimic estrogens and, in minute doses, cause a variety of reproductive problems, including infertility; some others cause dysfunction of the immune system or nervous system. The implication of this research is that water quality standards for synthetic organic pollutants provide inadequate protection for aquatic ecosystems, not only because they generally are not stringent enough but also because they address only a small fraction of the synthetic chemicals in the environment.[12]

Environmental standards based on concentrations in the water are the most common and probably the least effective way to regulate toxic chemicals in most marine environments. Substances introduced into the marine environment are mixed into the water and spread around by currents, so standards applying to the water become effective only once the entire circulation area (be it an estuary, a sea, or a large marine ecosystem) has become contaminated up to the numerical standard.

In addition, water quality standards are not applied to the parts of the marine environment where contamination by organic chemicals and toxic metals is greatest—the sediments and the microlayer. Some countries have sediment quality standards, which certainly protect benthic communities better than do water quality standards. However, there is still a great deal of controversy over how these standards should be determined and how much of the contamination in sediments is biochemically available and therefore a threat to the living community. Calculations based on the assumption that the only exposure route is through contact with the water in and above the sediments are the least protective. More realistic standards recognize the potential for ingestion by animals feeding in the sediments, absorption through membranes in direct contact with sediment, and bioaccumulation and biomagnification through the food chain.

Regulation related to water quality standards works by requiring permits for potentially polluting activities. Generally, for an activity to be permitted, it must be demonstrated that water quality or sediment quality standards will not be exceeded as a result of the activity. Therefore, this method of regulation allows polluting activities, unless otherwise prohibited, up to the point at which the water or sediments are contaminated at the levels of the standards. There is no provision to maintain the quality of less contaminated waters.

In addition to, or instead of, environmental quality standards, pollutants can be regulated by direct limits on emissions from identifiable point sources, such as industrial stacks and pipes, sewage treatment plants, and incinerators. Depending on their stringency, these regulations can enforce cleaner industrial practices and technologies, even in areas that are not immediately threatened with pollution. In fact, it is also possible to regulate the production technologies and methods themselves. Generally speaking, the closer to the source the regulations apply, the more effective they can be in minimizing contamination—depending, of course, on how rigorous and well enforced they are.[13]

Diffuse sources of toxic contaminants, called nonpoint sources, are more difficult to regulate directly. Contributions include pesticides from agriculture, silviculture (forest management), and lawn runoff and numerous substances in urban storm runoff. Generally, these sources are addressed by special programs designed to reduce contamination in the runoff. For example, crop-monitoring programs can reduce pesticide application by determining when it is really needed; most agricultural pesticides are applied on a regular schedule whether needed or not.

Environmental audits may be very effective in identifying both point and nonpoint sources of contamination into river basins. Such a technique was used for the Rhine River in Europe, which has a long history of delivering enormous loads of contamination from the several countries through which it passes on its way to the North Sea. The audit was initiated by the port of Rotterdam, which, because of the high levels of pollution coming down the river, could no longer operate without violating environmental laws and regulations. The audit identified sources of pollution; then it was necessary for the involved countries to agree on a plan (the "Rhine Plan") to reduce the sources of contamination. It will be a long process.

One of the countries through which the Rhine flows is Germany. With its many chemical and pharmaceutical companies, Germany is a major contributor to the problems of the Rhine. However, it is also an international leader in the development of clean production technologies, which are designed to reduce sources of contamination by improving production methods and products.

Clean Production Technology

One method of environmental protection that involves private industry as well as government regulations is the development and implementation of clean production technologies. Although the term *clean* refers primarily to the elimination of pollution, the concept also includes reduced consumption of resources. The philosophy underlying this concept is that industry can and should be nonpolluting, preserve biodiversity, and support the ability of future generations to meet their needs. In practice, clean production is really "cleaner production" and applies to industries that are making significant strides toward responsible use and reuse of resources and prevention of pollution of the environment. Such industrial systems seriously reduce or eliminate the use of toxic raw materials and avoid or significantly reduce the generation of toxic wastes and toxic products.

Criteria for clean production methods include (1) prevention of release into the environment of substances or energy that may cause harm to humans or the environment; (2) efficient and conservative use of energy; (3) use of renewable materials that are routinely replenished and extracted in a manner that preserves the viability of the ecosystem from which they are taken; and (4) incorporation of clean production standards into the entire life cycle of the product and its production. This "cradle-to-grave" concept begins with the selection, extraction, and processing of raw materials and ends with the recycling, reuse, or disposal of the product when it no longer serves a useful function. In between, it includes the design, manufacture, and use of the product.

An example of cleaner production that is being developed in the United States is the production of carpeting in which nontoxic materials are used, the carpet is manufactured in an energy efficient process, and the carpet is sold to a customer with a lifetime contract. The carpet is laid down without toxic adhesives, and when it needs replacing, it is taken up and reprocessed to a new condition. When old carpeting is taken up, the customer receives new (reprocessed) carpeting. The manufacturer remains involved, the use of new raw materials is greatly reduced, and the customer always has a presentable carpet.

The effect of clean production technologies on the marine environment is obvious. Reduction of toxic emissions from manufacturing, reduction of toxic runoff from landfills, and energy efficient design of processes ultimately reduce toxic pollution in the ocean. Reuse and recycling of raw materials protect the environments from which they originate.

Risk Assessment and the Precautionary Principle

Regulation of activities that affect the environment, including activities causing pollution and affecting fisheries, may be based on risk assessment, a process based on predicting the consequences of an activity. Estimated adverse effects are often balanced against known benefits of an activity. The best risk assessments attempt to deal with uncertainties associated with the predictions, but often these are ignored. The choice of end points in the risk assessment influences the estimation of risk, since less sensitive end points, such as death, will mask more sensitive end points, such as reduced reproductive success. The most sensitive end points, such as behavioral and developmental abnormalities, are not traditionally used but would provide more realistic assessments of risk. Nevertheless, the more sensitive the end point, the more uncertainty is associated with it.[14]

One tool used in risk analysis of pollution is the laboratory bioassay, in which a test organism is exposed to the pollutant or other stress in question. The simplest bioassays use one species and measure the percentage of deaths as the end point. However, if regulators were honest about it, they would have to acknowledge that there are serious problems with extrapolating the results of a bioassay at one level of biological organization to natural ecosystems, which reflect a far more complex level of biological organization. Bioassays involving several interacting species and employing more sensitive end points are more applicable to the real world. However, regulators don't like complex bioassays because they are more expensive and more difficult to perform properly and consistently. In the marine environment, the choice of species for these evaluations will also influence the outcome, even when the more sensitive sublethal end points are applied.

Ecologists are often technical advisors in decisions regarding environmental protection and sustainable use of resources. Ecological research, based on field studies, experiments, and modeling to predict ecological consequences of particular activities, is applied to the decision-making process with mixed results. Ecological predictions or hypotheses select the most likely outcome, generally ignoring the uncertainties in the knowledge on which they are based; hypotheses may change over time as more knowledge is acquired. Many surprising outcomes arise when the predictions fail.

The difficulties associated with selecting methodologies in risk assessments and dealing with associated uncertainties are rarely acknowledged, and this decision-making approach usually falls short of providing effective protection for the environment. The decision of what information to use in predicting environmental consequences of human activities becomes a question of policy. The adequacy of evidence used to make decisions regarding environmental protection should be, but rarely is, judged on scientific and ethical grounds. Doing so would lead to a different, risk-based decision-making process that takes uncertainty into account; assesses the risk associated with all possible outcomes, not just the most likely outcome; and judges acceptability of risk on ethical and scientific above economic and political grounds.[15]

One such alternative has become widely accepted in principle, though rarely in practice, in international agreements regarding marine environmental protection. It is called the *precautionary principle,* and it incorporates an ethical element. This approach emphasizes prevention of ecological damage by dictating that an activity be restricted or prohibited or that corrective action be taken before there is evidence of damage but when a significant negative environmental effect is deemed a strong possibility, even when the scientific evidence is inconclusive. Although on the surface this may seem to be a rejection of the idea of acceptable risk, it is actually near one end of a range of criteria for assessing risk in order to reach a decision that leads to the desired outcome. In the case of the precautionary approach, scientific and ethical criteria are applied to the assessment of risk and the protection of ecosystem integrity.

Monitoring and Research

Environmental monitoring is the measurement of environmental parameters at regular intervals over extended periods of time. Monitoring allows the assessment of environmental and biological changes in an ecosystem, with the goal of distinguishing natural fluctuations from abnormal changes as well as identifying cause-and-effect relationships between changes in the environment and changes in the biological community. Monitoring is essential to measure the progress and effectiveness of conservation measures, but even more valuable is the monitoring of marine systems over very long periods of time to detect biological trends in response to natural and human-induced perturbations in the environment.

Comprehensive monitoring programs are an important component of the assessment and protection of marine biodiversity. If we are ever to know the status of populations and species, we must have monitoring programs that assess them over large areas and long time periods. There have been very few long-term monitoring programs for the ocean, but those few have provided critical insights into the status of and changes in marine biodiversity. One such program is the CalCOFI (California Cooperative Oceanic Fisheries Investigations) program, which has been monitoring the physics, chemistry, and biology of the California Current system since 1949. These studies have yielded a wealth of information about global climate change; natural and unnatural fluctuations in populations of phytoplankton, zooplankton, fish, and planktonic larvae; biodiversity; and other important parameters. The long time frame of the studies has permitted assessments of cause and effect that are not possible in studies conducted over shorter time periods—for most research programs, a year to a few years. Other valuable long-term studies include phytoplankton records from Narragansett Bay in Rhode Island and plankton records from the North Sea. Both of these have provided and continue to provide invaluable insights into plankton biodiversity dynamics. However, funding for the long-term existence of some monitoring programs is uncertain, and there has been concern expressed that the "alarming termination rate of long-term monitoring programs in Europe is hindering the detection of ecosystem change in the ocean."[16]

Continued research on the relationship between living communities and their habitats is essential to improving our understanding of the dynamics of biodiversity in marine ecosystems. There is so much we do not know. Short-term studies and field experiments provide valuable information about species interactions that long-term monitoring cannot. Research and monitoring, however, are not a substitute for conservation programs; nor should waiting for the results of monitoring be an excuse for delaying action to protect habitats and biodiversity.

Economic Devices and Institutions

Economics is almost always a factor in policy decisions. Even when an environmental law or international agreement does not, in principle, incorporate economic considerations, its implementation often involves economic decisions. "Risk-benefit" analyses often become "cost-benefit" analyses, in which both sides of the analysis are in monetary units. The following discussion describes the various ways in which economics can be used to improve protection of the environment. The role of moral and ethical values is discussed here and, at greater length, in Chapter 8.

Environmental Accounting and Valuation of Living Creatures

The environment often loses in cost-benefit analyses for a couple of reasons: it is often impossible to assign a monetary value to an organism or an ecosystem, and the accounting assesses neither the cost of repairing an ecosystem that is degraded by the activity in question nor the benefits the undegraded ecosystem provides. Laws may override the economic assessment by requiring environmental protection at a cost or loss to an industry or individual. If fines are the enforcement tool of such a law, however, the analysis may still determine that it is more cost-effective to break the law and pay the fine than to protect the environment, to which no intrinsic value is attributed. At best, the accounting may assess the monetary value of resources that would no longer be available for the taking if they are damaged by a proposed activity; but if the activity is based on the taking of those resources (e.g., fishing and logging), the monetary value of the destroyed resource is counted as a benefit, not a cost.

The potential problems that arise when living resources become valuable only on their death is clearly demonstrated in a classic economic study of whaling in 1973, which demonstrated that the best economic strategy for hunting whales was to kill them all as quickly as possible and put the money in the bank. The reasoning was that the interest on the investment would accrue more rapidly than money could be made by sustainable hunting of whales at a rate that allowed reproductive replenishment of populations. Fortunately, the prescription was not followed to its end because other economic factors and international pressure intervened. Whaling more or less ended when moral outcry led to a global moratorium on the hunting of threatened and endangered whales. Although a few countries refused to obey the ban, the market for whale meat and products was greatly reduced. Even given the moral issue, the moratorium on whaling was agreed on in part for economic reasons—the industry had reached the point of limited returns because the increased effort and technology required to hunt the remaining whales was expensive.[17]

With accounting that evaluates a resource only by its monetary worth in the marketplace (usually dead and certainly out of its natural habitat), the only way ecosystems can win is if there is an established nondestructive marketable use that is more valuable in monetary terms than a proposed destructive use. With this in mind, some "ecological economists" have begun an attempt to estimate the annual value of "services" provided by the earth's living ecosystems. Market-based values are assigned on the basis of what it would cost to replace these services by technology. Previously, nature's services have been assumed to be free of charge and therefore without value, but this new approach makes it clear that without those essential services, a great deal of money would have to be spent to replicate them, assuming they could in fact be replicated. In a noteworthy attempt to determine the value of all the earth's living ecosystems, nature's services were estimated to be worth a minimum of $33 trillion, an amount that equals or exceeds the global gross national product. According to these estimates, terrestrial ecosystems were far exceeded by marine ecosystems, which were valued at about $22.5 trillion. The open ocean was valued at about $8.5 trillion, coastal marine ecosystems at about $12.5 trillion, and coastal wetlands at about $1.5 trillion. These high numbers apparently exceeded all expectations and are viewed by some as a wake-up call and an important guideline for environmental decision makers at local, national, and international levels.[18]

However, there is also a disturbing side to this economic analysis. Ignoring for a moment the moral dilemmas in assigning monetary values to living systems, there are some glaring omissions and questionable values. For one, the open ocean was assigned no value in climate control, though the question was left open for future analysis. Nevertheless, it is a surprising omission, considering the level of public awareness of the El Niño phenomenon, the effect of the ocean on global climate, and the common scientific knowledge that the ocean is responsible for a climate that can support life. An apparent ignorance of any value that the ocean has relative to our water supply is also disturbing, given the fact that replenishment of aquifers by rain is dependent on evaporation of water from the ocean. One wonders what the valuation of the open ocean would have been had just these two services been included.

On the other hand, some of the inclusions also raise serious questions. For example, waste assimilation or treatment by wetlands is given a very high service value. Yet this is not a natural service, since the toxic wastes trapped by these systems are unnatural products of human technology. It is true that wetlands naturally trap sediments and assimilate nutrients, but the trapping of human-generated wastes, especially toxic wastes, damages wetlands and diminishes other values of these systems. It is questionable whether this should be assigned a positive value, which might be used to validate the disposal of wastes into wetlands.

Clearly, this approach should be used with great care in conducting cost-benefit analyses for particular activities, since with different market conditions and different environmental conditions, the valuation could work quite differently. One real-life example involves a watershed in the Catskill Mountains that provides high-quality water for the city of New York. When it became apparent that the watershed was being threatened by development, the city conducted a cost-benefit analysis in which the cost of buying property to protect the watershed was compared with the cost of constructing and operating a water treatment plant to make degraded water from a degraded watershed potable. Protecting the watershed was determined to be much more economical, but at some time in the future this could change if the land becomes far more valuable as developable real estate and the technology for water treatment becomes relatively cheaper.

This system of economic valuation may not treat endangered species very well. The more rare a species becomes, the higher will be its value dead or captured (i.e., marketed) and the lower will be its value alive in the ecosystem, where it no longer plays a major role. The economics would favor letting endangered species go extinct. The good side of that argument, however, is that species would be valued more highly before they become endangered than they are under current environmental protection approaches.

The moral dilemma posed by the valuation of nature's services has given rise to some harsh criticism. Because of all the things nature provides that have no market equivalent or are not even known, such a system will always undervalue nature. Some observers also fear that it will promote the attitude that anything in nature is expendable—for a price. It may be that this new cooperative endeavor between science and economics merely puts off the difficult moral and ethical decisions that have to be made about the relationship of humans to nature. The researchers themselves recommend that moral values continue to be taken into account and incorporated into the valuation process. Indeed, some scientists find this economic approach appealing because it is quantitative.

Reward and Punishment

Other economic approaches to environmental protection involve rewards and/or punishments, otherwise known as incentives and disincentives. The use of such an approach requires that a set of guidelines be established for environmental protection. These may be binding obligations, such as laws that regulate or prohibit certain activities, or nonbinding suggestions for improvements in the ways in which activities are carried out. Once the guidelines are established, economic disincentives or incentives may be assigned to encourage compliance.

One disincentive that has been broadly recommended and sometimes adopted is the "polluter pays" principle. Its simplest application is the assessment of fines on a polluting industry that exceeds its permitted emissions levels, which are determined by a set of regulations. Polluters may also be held liable for the cleanup of pollution that they created in the past that still persists and for accidental spills. Sometimes the concept is formalized into a system of pollution rights that can be purchased. The way this works—in an estuary, for example—is that a total allowable annual input of pollutants is established and each contributing industry is given a share of that total as its annual permitted emissions limit. An industry that is designed or upgraded to emit less than its allotment can sell the unneeded pollution rights to another, dirtier industry. This provides an economic stimulus to install pollution prevention technology and pay for it by selling rights to pollute. To some, this may seem contradictory in principle, but others take the more pragmatic view that, while it does not reduce the allowable pollution, it may encourage conformity to the prescribed limits of pollution. Pollution is not the only type of activity for which economic disincentives may effect environmentally protective change. Resource exploitation, such as fishing, may also be subject to a set of rules and fines for breaking those rules.

Incentives or financial rewards include subsidies for those who comply with environmental protection guidelines. For example, subsidies could be given to fishers who adopt low-technology fishing methods, in contrast to the current system of subsidies that enable fishers to purchase high-tech fishing gear. Similarly, farmers could receive subsidies for not using pesticides, using less fertilizer, or adopting organic growing methods. In fact, there are programs that provide grants to farmers who install septic tanks for processing animal wastes into fertilizer and who establish buffer zones between their fields and streams.

Restoration and Mitigation

Although it is most desirable to protect coastal marine ecosystems in their natural condition, many have already been damaged or will be damaged before they are protected. For these ecosystems, restoration may be an option. Before living organisms are restored, however, the physical and chemical aspects of the ecosystem must be restored to a state that can support the desired community. This may range from the removal of interfering structures such as causeways, dams, and jetties to the dredging of contaminated sediments to the construction of reefs or the building of nesting platforms for endangered coastal birds. Recovery efforts may focus on single species or entire ecosystems. The latter, however, is usually accomplished by introducing new populations of the species that provide the ecosystem's fundamental structure and food chain base, such as salt marsh grasses, seagrasses, or corals, and hoping that the rest of the community will replenish itself. Such efforts have met with limited success.

Coastal wetland reconstruction has been the focus of a great deal of research and continues to be a goal. Although reconstruction of wetlands has proven to be partially effective, restoration of their full biodiversity and functioning has not been possible. This brings into question the wisdom of policies that allow destruction of wetlands for development purposes as long as equal acreage of new wetlands is created or restored. Obviously, protection and conservation of natural habitats and biodiversity are strongly preferred.[19]

Nevertheless, there are many degraded coastal ecosystems that would benefit from restoration efforts. Many, if not most, estuaries contain degraded habitat or are degraded throughout. Restoring them represents a monumental effort that can be attacked only a little at a time. Restoration of normal freshwater inflow is often one of the first objectives, but in areas where there are high agricultural demands on the water, this may be difficult. Strict pollution controls can initiate a long-term process of restoring clean sediments, but the controls must be faithfully employed along the entire reaches of inflowing river basins, where new sediments originate. In the United States, there are several regional estuary programs aimed at restoration as well as protection of their respective estuaries, including Chesapeake Bay, San Francisco Bay, Puget Sound, and others.

Disturbed coastal marine ecosystems are becoming more common, perhaps are even the norm, and may be accompanied by species extinctions or endangerment. Even if the stresses causing a given disturbance can be removed or greatly reduced, the system may not return to a desirable, productive, and diverse state. Even with the ability to intervene, it is often difficult to accurately predict the outcome of managed or natural recovery. The re-created ecosystem may be notably different from the original. Furthermore, restorers are not always aware of the natural condition that should be the goal of restoration, or they may not be able to return the system to its normal state because of physical changes that have occurred. Bottom trawling, for instance, kills off much of the benthic community and reduces the complexity of the sea-floor environment, thus reducing the area's potential species diversity and the community's ability to recover to its predisturbance character. The ecosystems that have been damaged need help, but there are limits to our capacity to restore natural marine ecosystems. Preventing damage is a much safer approach. As John Cairns stated in 1989, "If we are to diminish the global rate of species extinction and rehabilitate those species remaining, society, worldwide, must preserve the good condition of its remaining ecosystems."[20]

Chapter Seven

International and National Initiatives for Conservation of Marine Biodiversity

The rapidly changing conditions of the global environment pose threats to natural systems, human health, and national economies worldwide. The sources of the problems and their solutions can rarely be addressed effectively on a strictly national level. Members of the international community have, since the 1960s, slowly come to an awareness that they have a common interest in the conservation of wild species and the habitats in which they live. International cooperation is essential to protection of the environment and biodiversity of the seas. The marine environment transcends national borders, most of it is beyond the reach of any national authority, and activities on land or at sea in one country can often have far-reaching effects in other countries at long distances from the site of the activity. Laws and regulations at the national level implement international agreements and sometimes provide additional protection for coastal ecosystems in individual countries.

However, marine environmental protection has taken a long time to reach the negotiating table of the international community, and it has arrived there on three tracks. One track has been laid by the need for rules guiding the shared use of the seas for commerce. Throughout the long history of international trade by ships, it has been essential that the high seas be protected from claims of sovereignty in order to ensure free passage of ships bearing goods between countries. Another track deals with sovereign rights to resources within national ocean waters and the common rights of countries to marine resources outside national boundaries, that is, in the global commons. There has long been an interest in guaranteeing the freedom to fish the high seas, and nations began extending sovereignty offshore when it became technically possible to exploit the nonliving resources of the sea—oil and gas beneath the sea bed and hard minerals on the ocean floor. Only after exploitation rights were established did environmental restraints become part of the picture. The third track, therefore, deals directly with environmental protection and conservation of species. A consensus slowly arose to establish international rules that would balance sovereign rights over natural resources with the global need to conserve the natural environment and living species.

The shipping track led ultimately to the Convention on the Prevention of Marine Pollution by Dumping of Wastes and Other Matter, or the London Dumping Convention, in 1972 (now called the London Convention, 1972, or LC72), and the International Convention for the Prevention of Pollution from Ships, better known as MARPOL, in 1973 (now called MARPOL 73/78), both of which are evolving to be more prohibitive with regard to disposal of wastes at sea. The resources track led to agreements concerning marine mammal hunting, fishing on the high seas, international governance of Antarctica and its ice shelves, and the law of the sea, which gradually shifted from a focus on exploitation toward one on conservation. In 1958, the first United Nations Conference on the Law of the Sea was convened, and it led to several limited treaties. Then, in the early 1970s, negotiations began on the current comprehensive United Nations Convention on the Law of the Sea (UNCLOS), which was finally signed in 1982. It entered into force in 1994, after being ratified by sixty-eight countries. The natural ecosystem and species conservation track led to such international treaties as the Convention on International Trade in Endangered Species of Wild Fauna and Flora (CITES), the Convention on Biological Diversity, and many regional agreements.

One of the many accomplishments of UNCLOS is a codification of jurisdiction over defined areas or resources of the ocean. The convention declares all waters and sea beds between the influence of high tide and a distance of 12 nautical miles (about 13.8 statute miles) from shore to be publicly administered by the national government, free of private ownership. Nations also have rights to resources on the continental shelf and in waters out to 200 nautical miles (about 230 statute miles) offshore. Beyond that, the "high seas" are out of bounds of national jurisdiction and internationally agreed rules prevail. This scheme not only establishes zones where international law prevails but also dictates that all coastal waters will be publicly owned. The jurisdictional framework detailed in this convention provides the basic legal structure for other international treaties and national laws governing aspects of the marine environment and human activities in it.

Keeping coastal waters in the public domain facilitates coordinated environmental protection within and between countries that are so inclined. The shore, however, is not part of this convention. Thus, management and protection of coastal lands are complicated by private ownership, except in those countries or states that have individually extended public domain ashore to include all coastal lands under the influence of the sea, such as beaches, dunes, cliffs, and salt marshes. Some countries have taken that approach; others identify protected parks and refuges or protect particular habitat types.[1]

A medley of approaches have been used to protect biodiversity—protection of species, protection of special areas and ecosystems, regulation of human activities that threaten biodiversity, regulation of trade, and creation of sustainable development policies. Several international treaties, agreements, programs, and institutions employ one or more of these approaches. The conservation of biodiversity is the purpose of many, though not all, marine protected areas, and marine ecosystems have been included in instruments and programs developed to protect global biodiversity in general. Few international and national initiatives to protect the marine environment from particular human activities specify the use of biodiversity as a criterion for determining whether an area is being adequately protected.

The Process

Nations officially cooperate with one another on environmental matters through several formal institutions and instruments. Among these are the United Nations and a number of smaller institutions; treaties and other agreements, which may be bilateral, regional, or global; and a number of trade and financial institutions, such as the multilateral development banks, which lend or grant funds that have been supplied by richer countries for development projects in poorer countries.

The United Nations can afford protection to the marine environment through various programs and organizations, through global treaties negotiated under its auspices, through resolutions passed by the General Assembly, or through recommendations and agendas put forward by special commissions, conferences, and committees. International treaties are legally binding and generally take years to negotiate. They may involve only two or several nations, if the negotiations center on a jurisdictional dispute or on the environment of a regional sea area, or they may be global, open to all nations that wish to participate. In addition, there are other, nonbinding agreements among nations that are more rapidly negotiated. Although adherence is voluntary, these agreements express a consensus opinion that can have moral value in guiding the behavior of governments with respect to particular issues. These declarations of principle, resolutions, or charters may be quite effective in guiding actions, particularly when they are narrowly focused and when noncompliance will cause embarrassment for a government.

The process of negotiating a global treaty is often initiated by a conference on a topic or issue recognized to be of international importance and requiring some level of international consensus. It may be a ministerial-level meeting to which interested countries send representatives from the ministries responsible for the issues being discussed (e.g., appropriate departments and agencies in the United States). The conference may in some cases succeed in negotiating a treaty; more often, however, the result is a comprehensive agreement on the issues of the conference, possibly with a commitment to negotiate one or more relevant treaties. The Convention on Biological Diversity was adopted and opened for signature at the United Nations Conference on Environment and Development (UNCED, also known as the Earth Summit) in Rio de Janeiro, Brazil, in 1992, and the negotiations toward the adoption in 2000 of a treaty on pollution of the marine environment by persistent organic pollutants are the direct result of the Conference on the Protection of the Marine Environment from Land-Based Activities, convened in Washington, D.C., in 1995 by the United Nations Environment Programme (UNEP).

The negotiation of a treaty involves a series of meetings among different levels of government officials of the potential parties, that is, all those countries interested in playing a role in formulation of the treaty with the intent of becoming a party to it. Frequently, there are associated meetings of experts from interested countries, who meet in order to provide technical advice and recommendations for the treaty. Throughout the series of meetings, the treaty is drafted and revised to accommodate the interests and demands of all parties. Organization of the meetings and handling of documents are carried out through a secretariat, which is often housed in an international institution, such as UNEP or the International Maritime Organization.

The resulting treaty usually includes several compromises, concluded after discussions among parties with different opinions on particular provisions of the treaty. Sometimes no compromise can be reached and disputed items are dropped or carried over with the intent of subsequent revisions. In the end, when a final draft is agreed on by consensus or majority, it is signed by all participating parties who can agree to it in principle. This document is then taken back to the respective governments for ratification, a process that may take many years. The United Nations Convention on the Law of the Sea, for instance, took ten years to negotiate to the signing stage and another twelve years to be ratified by enough countries to bring it into force, thereby becoming legally binding for all ratifying countries. In order to get it ratified, further negotiations were necessary on the section governing exploitation of minerals on the sea bed. Once a global treaty is ratified and goes into force, it remains open for any additional interested countries to ratify, whether they were initial signatories or not.

A treaty or convention is commonly an instrument that changes and develops over time. Frequent (often annual) meetings of committees of experts and of the signatory parties provide opportunities for developing guidelines, discussing problems, and monitoring progress on implementation of various provisions of the treaty. The parties can change a treaty through a specified amendment process, which requires consensus or a majority vote. Sometimes, separate protocols are negotiated within the context of a convention in order to add provisions consistent with but not included in the original agreement.

During negotiations and subsequently, during annual meetings, international nongovernmental organizations (NGOs) may participate as observers. NGOs are special interest groups such as international environmental, professional, and industrial organizations. To participate in treaty negotiations, an NGO must first apply for observer status and be approved by the parties to the treaty. Treaties vary considerably in the difficulty of obtaining observer status and in the rules observers must follow at meetings. Although observers never have votes, they are sometimes allowed to speak briefly during meetings, and they can generally submit formal or informal papers pertaining to issues on the agenda, which are distributed to the official delegates. NGO observers also use the opportunity to discuss the issues informally with delegates outside the meeting room.

Once a country has ratified a treaty, it is obliged to establish a national legal framework to implement and enforce the treaty's provisions. In the United States, congressional approval is required for ratification, and then it is the responsibility of Congress to pass the appropriate laws. Once the laws are in place, the responsible agencies must develop detailed regulations to provide a means of implementing and enforcing them. National implementing laws must not be weaker than the convention, but they usually may be stronger as long as they do not infringe on rights guaranteed to other countries by the treaty. A similar hierarchy exists within the United States, where state environmental laws can be stronger, but not weaker, than the corresponding national laws.

Sometimes national laws and regional treaties guide global treaty negotiations. The United States was a leader in environmental legislation in the 1970s and 1980s, and some international treaties are patterned after U.S. initiatives. Whether this set the stage for the current posture of the United States in international negotiation or whether it is due more to this country's sense of power, the U.S. government is very reluctant to agree to any provisions in treaties that exceed domestic laws and practices. Sometimes, however, the combined pressure from other countries and from citizens within the United States is great enough to effect a change in policy.

In some areas of environmental protection, Europe has become the new leader, with some regional agreements, such as those concerning the North Sea environment, setting the stage for global agreements. Furthermore, some industrialists in Europe and Japan have taken the call for environmental protection as a challenge to develop and market technologies and processes that are more environmentally sensible, whereas most—though certainly not all—U.S. industries lag behind in that respect, continuing to object to stringent regulations and clinging to wasteful and polluting processes.

It is important but often difficult for citizens to become informed about international environmental treaties that their governments have ratified and the purposes behind the laws that implement them. They should also be aware of those treaties that their government has refused to ratify and why. Environmental NGOs provide one link between citizens and the negotiation and implementation of international agreements; in the United States, Congress should provide another, but knowledge and involvement vary greatly among congressional officials. Few citizens in the United States express great interest unless extensive advertising campaigns are waged in the media by industries that feel threatened and by environmental organizations that believe the stakes are high. In fact, the news media are probably the most important sources of information for the public, so the media's interest or lack of interest in a treaty might well be reflected in the general population. To promote concern about treaties and involvement in their implementation, the press must become engaged enough to engage the people in turn. Similarly, trade treaties and international lending institutions should be monitored by NGOs, news media, and citizens with respect to the environmental effects of their requirements and policies and the role of the United States government in setting those policies.

Within the United States, the agencies most often responsible for marine environmental regulations and programs that implement related laws and treaties are the Environmental Protection Agency (EPA), the National Oceanic and Atmospheric Administration (NOAA), the U.S. Army Corps of Engineers (COE), and the United States Coast Guard (USCG). The National Marine Sanctuary Program, administered by NOAA, is the only program with the specific goal of protecting marine biodiversity. NOAA's National Sea Grant College Program promotes applied coastal research at universities, and other research programs such as the National Science Foundation, the Polar Research Board (part of the National Research Council), and the Office of Naval Research, support important marine research. None of these focus on environmental protection per se, but they supply funding for research that provides information critical to decisions about the environment.

Internationally, several institutions administer programs that are relevant to the conservation of marine biodiversity. Some of these are arms of the United Nations that have responsibility for international affairs, including certain aspects of the marine environment. The United Nations Environment Programme (UNEP) is the most directly involved in conservation, but others, such as the Food and Agriculture Organization of the United Nations (FAO), which oversees world fisheries, and the United

Nations Commission on Sustainable Development, are also important. Other programs and organizations have been created by treaties to administer or provide advice regarding their respective treaties and their implementation. The International Whaling Commission (IWC) is one, and the Joint Group of Experts on the Scientific Aspects of Marine Environmental Protection (GESAMP) is another. The International Maritime Organization (IMO), which governs international ship commerce, is noteworthy for its success and efficiency in carrying out its mission.

Yet other programs have been created to promote and finance development projects in developing countries. Multilateral development banks, for instance, are institutions that facilitate the transfer of funds, generally as loans, from a pool of money provided by wealthy member nations to poorer nations wanting to raise their standard of living and stimulate their economic growth. Although these institutions initially had no interest in environmental matters, their negative effect on the ecosystems of borrowing countries became so great that they were shamed into a concern about environmental protection. Some of these, such as the World Bank, now have environment departments that are philosophically far beyond most governments in their goals for environmental protection. There are also a few independent organizations that promote conservation among governments. The World Conservation Union (formerly known as the International Union for the Conservation of Nature and Natural Resources, and still using the acronym IUCN), with both governments and nongovernmental organizations as members, is the most important of these.

Some of the more important initiatives and institutions that most directly pertain to the protection of marine biodiversity are briefly described on the following pages. There are numerous others that are relevant but cannot be covered in the space of this book. In many cases, these treaties, laws, and programs have not yet fulfilled their own aspirations; nor have they provided adequate protection for the world's ocean and its living communities. Nevertheless, they provide a framework within which significant progress can be made with relative ease if countries and citizens have the will to implement them. It is that potential as much as their accomplishments to date that makes these instruments so important. For more information on specific laws, treaties, programs, and organizations, the Internet is a valuable resource, as are the secretariats of the various conventions and organizations; IUCN and the International Whaling Commission publish compendiums and reviews of various conventions.[2]

The Ocean Environment

Because of the interconnected nature of the marine environment, the most effective way to protect marine biodiversity is through comprehensive protection of the ocean and its tributaries. In fact, the most effective approach would be the reverse of protected areas such that the whole ocean system would be protected from all human activities, with specially designated areas or lanes where particular activities would be permitted. Such permit areas would be carefully chosen to minimize adverse effects on the marine ecosystem as a whole while providing the desired services or resources. Since humans do not live in the ocean and therefore don't require its space to accommodate uncontrolled population growth, it would theoretically be possible to apply this reverse strategy in allocating ocean resources. But such an approach is hardly likely to be adopted, since it is so different from the customary way of doing business. Instead, there are agreements, laws, and programs that grant consideration to the marine environment and specify special protective measures for particular species or areas or prohibit or restrict particular activities. Few attempt to coordinate activities over the entire marine environment.

Law of the Sea

The United Nations Convention on the Law of the Sea (UNCLOS), which has eighty-one party nations (nations that have ratified, as of 1997), sets up the framework for allocating and regulating the taking of marine resources and the use and protection of the marine environment—that is, it establishes the rules for all human uses of the marine environment. It is the most comprehensive marine treaty and offers a framework within which marine environmental protection is a priority. However, its effective implementation will depend on the commitment of the party nations and their citizens.

Among this treaty's numerous accomplishments is the definition of areas of jurisdiction. *Internal waters* constitute the zone closest to shore and include all waters on the landward side of a *baseline,* which varies somewhat from one country to another but is generally the high-tide level on a coast or the upstream limit of marine influence, as determined by salinity, in a river mouth. This use of the word *baseline* bears no relation to the ecological baselines introduced in Chapter 3. The next zone outward from shore is the *territorial sea,* which nations may extend out to 12 nautical miles (about 13.8 statute miles) from the baseline. Both internal waters and territorial seas lie within national jurisdictions, but in territorial seas the right of "innocent passage" is guaranteed to ships from other countries. Beyond 12 nautical miles and extending to a maximum distance of 200 nautical miles (about 230 statute miles) from the baseline, is the *exclusive economic zone* (EEZ), which gives nations jurisdiction over fisheries and preservation of the marine environment but does not allow them to restrict the navigation of foreign vessels. Nations also have the sovereign right to exploit the natural resources of their own sections of continental shelf, with jurisdiction extending across its entire extent, even if it is wider than 200 nautical miles. Beyond all countries' EEZs lie the *high seas*, where no nation has jurisdiction and any regulation of resources or the ocean floor must be the subject of international agreement.

Although most of UNCLOS relates to the use and exploitation of the ocean and its resources, the treaty also contains a section on protection of the marine environment (Part XII, Articles 192–237). All nations are obligated "to protect and preserve the marine environment" (Article 192) and to take all necessary measures to prevent, reduce, and control pollution. This commitment to protect and prevent represents an early version of the precautionary principle. It is further expressed in the definition of pollution, which includes the anticipation of harm as well as harm done. Pollution includes products and by-products of technology and the introduction of exotic species capable of harming the marine environment. It includes pollution from sources on land as well as at sea and pollution throughout the atmosphere. Although the treaty does not invoke biodiversity per se, it does make it clear that harm to the marine environment would include damage to or loss of biodiversity. Furthermore, the convention encompasses to some degree the importance of species interdependencies and ecosystem considerations in the management of fisheries.

UNCLOS reaffirms the traditional right of all nations to fish the high seas but introduces qualifications for maintaining that right, including (1) the responsibility to protect straddling stocks and stocks that migrate beyond the EEZ and (2) the duty to cooperate in conserving and managing high-seas living resources, taking into consideration the best scientific data available. Within a coastal country's EEZ, that country is responsible for conserving and managing living resources to ensure that they are not endangered by overfishing. The standard adopted, however, is that of maximum sustainable yield, which many biologists do not consider protective. The deference to the best available scientific data characterizes all conservation responsibilities in the treaty.

The breakthrough that finally brought the convention into force after so many years was, as mentioned, an agreement to change the provisions regarding mining of the deep sea bed. By the time of this agreement, many of the countries that had resisted ratifying the convention had incorporated many of its other provisions into their laws and policies. The renegotiated section on sea-bed mining requires that marine resources and the environment be protected from the harmful effects of the activity and sets up an International Seabed Authority to regulate sea-bed minerals exploration and exploitation beyond the continental shelf and EEZ areas. This authority is charged with adopting rules, regulations, and procedures that will protect and preserve the marine environment and with promoting research and monitoring toward that end. Within national jurisdictions, the responsible governments are called on to do the same with rules no less effective than the international rules.

UNCLOS allows for the establishment of protected areas to prevent serious harm to the environment from mining of the deep sea bed and from shipping, with particular attention to ice-covered seas. Designation of such areas in relation to vessel passage is left to the International Maritime Organization (IMO). Species protection is afforded by the obligation of nations to "protect and preserve rare or fragile ecosystems as well as the habitat of depleted, threatened or endangered species and other forms of marine life" (Article 194.5).

Global Agendas

The United Nations began formally discussing the linkages between development and environment in 1972 at the United Nations Conference on the Human Environment, held in Stockholm, Sweden. The result of that meeting was the Stockholm Declaration, which recognized fundamental human rights to "freedom, equality and adequate conditions of life in an environment of a quality that permits a life of dignity and well-being" and proclaimed the responsibility of all governments to protect and improve the environment for both present and future generations. These ethical guidelines have persisted in principle in international negotiations on the environment since that time, and the concept of responsibility to provide for future generations is now commonplace. These guiding principles were carried into the United Nations Conference on Environment and Development (UNCED), also known as the Earth Summit, held in Rio de Janeiro, Brazil, in 1992. Implementing these principles is, of course, not as easy as speaking about them or affixing a signature to the words, and nations still argue about what needs to be done and how to accomplish it.

The Stockholm meeting led to the establishment of the World Commission on Environment and Development, which issued a report in 1987 titled *Our Common Future.* This report recognized the importance of the ocean environment in meeting the commitment to achieve a sustainable world:

> Looking to the next Century, the [World Commission on Environment and Development] is convinced that sustainable development, if not survival itself, depends on significant advances in the management of the oceans. Considerable changes will be required in our institutions and policies and more resources will have to be committed to oceans management.[3]

The negotiations for UNCLOS took place in the wake of the Stockholm Declaration, as did UNEP's initiative to develop guidelines for governments in addressing adverse effects on the marine environment caused by human activities on land. In 1985, the Montreal Guidelines for the Protection of the Marine Environment from Land-Based Sources of Pollution were adopted.

The 1992 conference in Rio de Janeiro (UNCED) resulted in the Rio Declaration, which sets down principles accepted by the signatory governments, and Agenda 21, which is a plan of action for addressing the issues raised at the conference, guided by the adopted principles. The precautionary principle, discussed in Chapters 6 and 8 of this book, is one of those. This set the stage for incorporating the precautionary principle into international environmental agreements that have been negotiated or amended since then. Agenda 21 has one chapter devoted to conservation of biodiversity (Chapter 15) and one that addresses the ocean environment (Chapter 17), including degradation of the marine environment by land-based activities.

Governments participating in UNCED committed themselves to applying preventive, precautionary, and anticipatory approaches to marine environmental protection; to conducting environmental impact assessments when proposed activities threaten to degrade the marine environment; to integrating marine environmental protection into development activities; to developing economic incentives for clean technologies and developing means to avoid degrading the marine environment; and to improving living standards in coastal communities. In fulfillment of obligations in Agenda 21, UNEP initiated a process that culminated in adoption of the Global Programme of Action for the Protection of the Marine Environment from Land-Based Activities at a conference in Washington, D.C., in 1995.[4]

The Global Programme of Action addresses all types of activities on land that threaten the marine environment and prescribes actions to be taken, along with their objectives at the national, regional, and international levels. An important provision of the Global Programme of Action is the commitment of governments to develop comprehensive national programs of action to address land-based activities that affect the marine environment. These national action plans are crucial if anything concrete is to be done; citizens and NGOs in all countries should be following this process closely and helping to guide their governments to make sure the plans are strong and effective. The conservation of marine biodiversity could be greatly enhanced by these initiatives.

U.S. Laws and Agencies

An act of Congress in 1966 established the Commission on Marine Science, Engineering and Resources, known as the Stratton Commission after the name of its chairman. The commission issued a report in 1969, *Our Nation and the Sea: A Plan for National Action,* which shaped U.S. ocean policy, with a focus on exploitation of ocean resources, for the ensuing three decades. In honor of the United Nations Year of the Ocean, 1998, a new act, known as the Ocean Act, established a new commission, the National Ocean Council, to revisit the issues and provide advice for development of a new national ocean and coastal policy that should provide for environmental protection and resource management as well as research and development.

The National Oceanic and Atmospheric Administration (NOAA) within the U.S. Department of Commerce, is the primary agency dealing with the marine environment. Its National Marine Fisheries Service is responsible for regulating marine fisheries and protecting marine mammals. It also houses the Marine Sanctuary Program, the National Estuarine Research Reserve System, and the National Sea Grant College Program, and it is responsible for conducting considerable research and monitoring in the marine environment and assessing the status of marine ecosystems. The Environmental Protection Agency administers most regulatory programs having to do with water quality, including marine water quality. The U.S. Department of the Treasury has adopted policy standards for lending money for projects that may have an effect on the marine environment, but it is not clear how carefully they are applied.

Biodiversity

Marine biodiversity is also offered protection under instruments and institutions that are concerned with biodiversity in general. There are no agreements or laws that apply directly and solely to marine biodiversity.

Convention on Biological Diversity

The Convention on Biological Diversity is a comprehensive treaty that doesn't treat marine ecosystems significantly differently from terrestrial ecosystems. It does, however, establish a framework within which parties to the convention can provide significant protection for all types of biodiversity in the marine environment—genetic, species, functional, and ecosystem diversity.

The main objectives of the convention are conservation of biodiversity, sustainable use of its components, and equitable sharing of the benefits arising from the utilization of genetic resources. The precautionary principle is adopted in the preamble, but there is little guidance for applying it, except for a statement that lack of scientific information is not a reason for delaying conservation action when it is expected that biodiversity is or would be threatened.

A major part of the convention deals with the transfer of genetic resource conservation technologies, such as gene banks, and genetic engineering technologies to developing countries to enable them to protect and retain the genetic resources native to their countries, as well as the guarantee of intellectual property rights, such as patents, over their national genetic resources. Although these concerns are not immediately applicable to the marine environment, that situation will change as aquaculture and other genetic engineering technologies are developed for marine creatures.

More applicable are the requirements for parties to develop national strategies, plans, or programs for the conservation and sustainable use of biodiversity and to integrate conservation and sustainable use of biodiversity into other policies and programs. There is particular emphasis on *in situ* conservation, the maintenance of viable populations of species in their natural surroundings. Parties to the convention have an obligation to prevent the introduction of exotic species; however, they are also required to eradicate or control those already introduced. The latter provision could backfire if risks are not assessed on a species-by-species basis, since many methods of eradication and control can threaten other species in the habitat. The convention also prescribes that risks associated with the use and release of living, genetically modified organisms be "regulated, managed and controlled." This is contrary to the precautionary principle, which would suggest that such risks be prevented; in the marine environment, where it is more difficult to monitor such releases, this may be a prescription for disaster.

The emphasis of the convention is on the conservation of ecosystems rather than species per se, except that legislation for protecting threatened species and populations is required. Aside from that, there is an obligation to identify and monitor species considered to be important elements of biodiversity, which apparently include threatened species and species of economic or other value for human use—an odd definition of importance to biodiversity and certainly in conflict with ecological principles. One provision that could be useful in regulating fishing or outlawing certain fishing technologies is the requirement to adopt measures to avoid or minimize adverse effects on biodiversity resulting from the use of biological resources.

The Convention on Biological Diversity obligates parties to prepare environmental assessments of proposed projects that are likely to have significant adverse effects on biodiversity. Unfortunately, this obligation can be avoided by the use of weak standards for determining the likelihood of significant adverse effects. Parties are also required to take biodiversity and environmental consequences into consideration in their programs and policies. All the conservation obligations are qualified by the phrase *as far as possible and as appropriate* to compensate for the different capabilities, both financial and technical, of different countries, especially developed and less developed countries. However, that phrase makes the conservation provisions weak and possibly ineffectual.

The effectiveness of this convention will depend on the commitment of national governments, to which all responsibility reverts; the effectiveness of oversight of implementation; and the provision of financial aid for its implementation in developing countries. It was adopted in 1992, but it has not yet been ratified, so it is not in force. The United States has not ratified the convention and has no comparable legislation, although it sends a delegation to the treaty meetings and actively participates in policy discussions and influences decisions.

Initiatives to Protect Endangered Species

Less effective in protecting biodiversity, but still important in the absence of comprehensive protection, are the various treaties, laws, and policies protecting individual species. Insofar as protecting species often means protecting the ecosystems that support them, these instruments can offer more far-reaching protection to biodiversity associated with those ecosystems, and that may be their major contribution, even though it is not their major objective. Most of these initiatives focus on saving endangered and threatened species from extinction, and as such, they apply only after the species has ceased to play a major role in its ecosystem. A "saved" species can therefore continue to contribute to species counts; however, it does not contribute to functional diversity unless or until it resumes its former population size and role in the ecosystem.

The Convention on International Trade in Endangered Species of Wild Fauna and Flora (CITES) is a global convention that has been implemented and enforced with some success by a large number of countries. The treaty is based on lists of species that are (1) endangered, and therefore no international trade is allowed; (2) threatened, and therefore trade is restricted and permits are required; or (3) listed by one of the parties to the convention as needing cooperation from other countries to prevent export from the listing country without a permit. The prohibitions or restrictions apply to trade that is believed to threaten the survival of a species. There are various exceptions that require special permits. This treaty has effectively reduced trade in many conspicuous terrestrial species whose populations are in trouble, but it has not been particularly helpful for most marine species because the status of these species is so difficult to determine. A few corals and other reef species have been listed, as have a few nonreef fish, sea turtles, seabirds, and marine mammals. However, the list of marine species is far from inclusive because of lack of information and because those that are subject to trade, especially commercial fishery species, often reach population levels that are not worth the extra fishing effort before they become endangered as defined by the treaty. Since CITES does not protect separate populations of species, its value in the marine realm is limited.[5]

Another treaty, the Global Treaty on Migratory Species, adopted in 1979, has a similar focus on endangered and threatened species and on the need for countries to cooperate in providing sufficient protection for species, both terrestrial and marine, that migrate between jurisdictions. This treaty covers numerous marine species that migrate between territorial seas and EEZs of different countries, but it does not cover the high seas. In a similar fashion to CITES, this treaty lists species in different categories, in this case endangered species and species likely to benefit from international cooperation. Marine mammals, sea turtles, seabirds, and fish are among the marine species listed. Parties are prohibited from taking animals on the endangered list and are obligated to conserve and restore the habitats of these species. They must also prevent, reduce, or control factors likely to further endanger these species, including the introduction of exotic species. The species on the list for international cooperation are to be protected by agreements among countries that share populations of any of the listed species, a provision useful in the marine environment.

The Endangered Species Act (ESA) of the United States has also been successful in protecting species and their habitats on land, and it is increasingly being applied to marine species. The U.S. Fish and Wildlife Service, in the U.S. Department of the Interior has primary responsibility for implementation of the ESA, with some share of responsibility going to the National Marine Fisheries Service of NOAA, in the Department of Commerce. The ESA authorizes the designation of habitats critical to the survival of listed species and directs the establishment of programs for the recovery of listed species. It also provides for cooperation with other countries to conserve endangered and threatened species and authorizes prohibitions and permit controls in the taking and trade of such species. Federal agencies are prohibited from funding, authorizing, or carrying out any projects that jeopardize the existence of or modify the habitats of endangered species.

This act is widely disliked by developers and those who exploit natural resources because it often gets in the way of their plans. It is the critical habitat provision that causes them the most trouble. The constant battle between environmental protection and exploitation typically reaches a crescendo during years when the ESA is up for reauthorization—a period when it can be amended and made stronger or weaker.

Initiatives to Prevent Species Introductions

Although there are no treaties dealing exclusively or principally with introduced species, there are several nonbinding international measures to reduce such introductions. Since ballast water is the primary source of marine species introductions, the Marine Environment Protection Committee of the International Maritime Organization (IMO) in 1991 developed International Guidelines for Preventing the Introduction of Unwanted Aquatic Organisms and Pathogens from Ships' Ballast Water and Sediment Discharges. Two other institutions, the International Council for the Exploration of the Sea (ICES) and the Food and Agriculture Organization of the United Nations (FAO), have issued codes of practice regarding the introduction of species. The ICES code addresses the risks of introducing or transferring any marine organisms, including aquaculture organisms, whereas the FAO code addresses the risks of introducing fish species.

Limits on Exploitation

Another way to protect biodiversity is to place limitations on exploitation activities in general or on the exploitation of certain types of organisms.

Initiatives to Protect Marine Mammals

The International Convention for the Regulation of Whaling, signed in 1946, established the International Whaling Commission (IWC), which regulates the hunting of whales by establishing bans, catch and size limits, and gear specifications. It also disseminates information on whale biology and ecology and on international initiatives that have direct or indirect relevance to whales and whaling. The original purpose of the IWC was to regulate the killing of whales for the benefit of the whaling industry. However, as membership was open to all countries, many nonwhaling countries joined and soon outnumbered whaling nations.

The initial efforts of the IWC were an abysmal failure, and the stocks of all the great whales continued to plummet despite catch limits, which had been placed at too high a level to be helpful and were grossly violated by countries such as the Soviet Union. Finally, nonwhaling countries outnumbered whaling countries by more than the two-thirds majority required to pass an indefinite moratorium on whaling; the moratorium took effect in 1986. Japan, Norway, and the Soviet Union, however, protested and continued whaling. Japan and Norway continue to this day, in some cases using loopholes in the moratorium to "legally" hunt. For instance, these two countries, claiming to hunt for scientific research (allowed under the moratorium), each kill more than 500 minke whales per year, which are in fact sold to wholesalers. Recent DNA studies have uncovered whale meat from several of the banned species in the Japanese marketplace, whereas Norway appears to be restricting its hunt to minkes, not considered endangered. Both countries have announced an intent to continue commercial whaling. Whaling by indigenous peoples for cultural and subsistence needs is also permitted and practiced in several Arctic regions and, most recently, by coastal tribes in the United States. Currently, forty countries, including Japan and Norway, are members of the IWC. Iceland has declined membership, asserting a right to hunt whales when it wishes.

A proposal to the IWC by Ireland in 1997 and still being debated in 1998 would legalize the hunting of whales but would establish a Revised Management Scheme, designate a global sanctuary for whales, allow closely regulated and monitored coastal whaling within national EEZs, prohibit international trade in whale products, and end scientific research kills. Supporters argue that strict oversight of legal hunting would actually reduce the whale kill, which has been rising steadily within the loopholes or in violation of the current moratorium. Detractors believe it would only invite countries that have abandoned whaling to start once again.

The taking of small cetaceans (dolphins and porpoises) is not regulated by the IWC, but the United States and other countries are pressuring the commission to do so. The issue is becoming urgent because of the large Japanese catch to supplement the whale meat market.

In the United States, marine mammals are protected by the Marine Mammal Protection Act (MMPA), and several species are also listed under the Endangered Species Act (ESA). The MMPA, enacted in 1972, recognizes the need to protect all marine mammal species. The National Marine Fisheries Service of NOAA manages whales, dolphins, porpoises, seals, and sea lions, whereas the U.S. Fish and Wildlife Service is responsible for sea otters, manatees, walruses, and polar bears. The primary management tool is a prohibition, with some exceptions, on the taking of marine mammals. The term *taking* means hunting, capturing, killing, or harassing. Special restrictive permits are authorized for scientific research, aquariums, subsistence purposes, and a limited number of incidental catches in association with commercial fisheries. The level of protection afforded a species depends on assessment of its status, and 1994 revisions to the act recognize uncertainty and provide for conservative estimations of missing information about populations and growth and recovery rates, which results in more restrictive management than in the past when the status of a species is uncertain. In addition to managing takings, the act prohibits most importation of marine mammals and their products.

Global Fisheries Initiatives

A United Nations treaty to regulate fishing on the high seas—officially the Agreement for the Implementation of the Provisions of the Convention on the Law of the Sea of December 1982 Relating to the Conservation and Management of Straddling Fish Stocks and Highly Migratory Fish Stocks—was negotiated as an outgrowth of Agenda 21 and opened for signing in December 1995. It will become legally binding when thirty countries have ratified it. The agreement aims to prevent overfishing and ease international tensions over competition for dwindling fish stocks. It requires countries to conserve and sustainably manage exploited fish populations and to peacefully settle disputes over fishing on the high seas.

Specifically, the treaty establishes a basis for sustainable management of the world's fisheries; addresses the problem of inadequate data; provides for establishment of quotas; calls for regional fishing organizations where none exist; addresses problems caused by the persistence of unauthorized fishing; establishes procedures for ensuring compliance; and prescribes options for settling disputes between countries. However, responsibility for regulating and enforcing sustainable fishing practices reverts to regional organizations. They are obliged to collect data on catches and allocate quotas for individual countries, and they can take enforcement action against any ship or boat that undermines the agreed conservation regime. The treaty adopts the precautionary approach in the face of incomplete information and uncertainty, but it does not prescribe particular regulatory actions or standards. That is left to the regional organizations. Therefore, these provisions will vary in effectiveness as the treaty is implemented, which is not expected to happen until 2002 or beyond.

Another United Nations initiative focusing on marine fisheries and protective of biological diversity is a 1989 nonbinding resolution, Resolution 44/225. It came about as a result of worldwide outrage that accompanied increasing public knowledge of the wanton damage drift nets were causing. The resolution recommends a moratorium on the use of large drift nets (many miles long) on the high seas beginning in July 1991. Although it is nonbinding, it carries the moral force of the United Nations General Assembly, and most countries with fishing fleets that used such nets have ended their use. The objectives of this resolution were accomplished much more quickly than they could have been through the treaty process, wherein negotiation and ratification usually take several years. National waters of the South Pacific are covered by a treaty that bans all nets longer than 2.5 kilometers (1.6 statute miles), and drift nets are banned in the North Pacific by three separate agreements collectively involving the United States, Japan, Canada, Taiwan, and Korea.

The Food and Agriculture Organization of the United Nations (FAO) monitors fish stocks and landings (the portions of catches kept and brought ashore) worldwide. The organization has compiled a valuable source of data to inform management strategies. The data, however, are far from complete and rely on accurate reporting from the fishing nations, among which the quality of data surely varies. Nevertheless, based on the data it has, the FAO has estimated that 70 percent of commercial fishing grounds are depleted or are recovering from overfishing.

Regional Agreements on Fisheries

Although a comprehensive discussion of the multitude of regional international environmental agreements is beyond the scope of this book, it is worth mentioning that fishery regulations seem to be more manageable on regional than global levels. Moreover, there seems to be greater motivation to develop regional agreements to regulate fisheries than there is to regulate on a strictly national level. Sometimes the agreements are aimed at dividing fishing grounds among national fleets, and sometimes they address the need for cooperation in conserving waning fish stocks. They generally focus on exploitation rather than on habitat.

Regional agreements have focused on a number of different species, usually prompted by disagreements among two or more nations regarding exploitation of the species in question. One of the most prominent such agreements, the Agreement to Reduce Dolphin Mortality in the Eastern Tropical Pacific Tuna Fishery, was adopted in 1992 under the auspices of the Inter-American Tropical Tuna Commission and formalized as the Panama Declaration in 1995. Since it would weaken the definition of dolphin safe tuna but would strengthen international cooperation, there are differences of opinion within the conservation community as to whether this agreement represents improved protection for dolphins. Atlantic tuna are managed by the International Commission for the Conservation of Atlantic Tunas. Other regional agreements focus on pollock in the Bering Sea, southern bluefin tuna, anadromous fish (salmon) in the North Pacific Ocean, northeastern Atlantic fisheries, and North Atlantic salmon.

Marine Fisheries Management in the United States

Fisheries in the United States are regulated by the National Marine Fisheries Service (NMFS) of NOAA. With the enactment of the Magnuson Fisheries Conservation and Management Act in 1976, U.S. jurisdiction and control over all marine fishery resources was extended to 200 nautical miles from the coastline, consistent with the provisions being negotiated for UNCLOS. The purpose of the action, which was taken largely in response to the increasing presence of foreign fleets, was to secure the exclusive rights of U.S. fishing fleets to exploit fish populations. The act established eight regional fishing councils charged with preparing, monitoring, and revising fishery management plans in the U.S. EEZ of that region. In practice, however, until recently very little management was done because the councils were composed entirely of representatives of the fishing industry who did not want to be regulated. Under this act, an estimated 42 percent of fishery species were depleted. One of its very few success stories is the recovery of the striped bass as a result of management in the northeastern region; far more common was the collapse of one fishery population after another.

In the face of these failures, it became apparent in the 1990s that the Magnuson Act needed revising and the regional councils needed reforming. In 1996, after several years of debate, the act was revised with passage of the Sustainable Fisheries Act. This act includes several key changes in the way marine fish are to be managed. The establishment of catch limits to prevent overfishing is commonplace in freshwater fisheries but was almost absent, before the revised act, in the management of marine fisheries. Previously, in the rare cases in which limits were imposed, they were more favorable to the industry than to the fish. With the new act, fishery management plans must include measures to restore populations of overfished species, and these populations must be assessed every year to determine whether the management regime is working or needs to be altered. Although these measures are intended to maintain viable fishery stocks, they can also protect biodiversity by reducing pressure on the ecosystems, including other species that support the commercial species.

The part of the act that more directly addresses biodiversity concerns is the part dealing with protection of fish habitat. Fishery management plans are supposed to minimize adverse effects on habitat due to fishing, which could have a significant effect on the types of gear used in some fisheries. The act also sets reduction of bycatch as an objective and thus protects a greater variety of species than those considered commercially valuable. However, there is concern that the ecosystem is still not being assessed and protected as a whole. The act does not address the problems with composition of the management councils, so their reform will have to come from within and may therefore be slow. Pressure is building to include members that represent perspectives and expertise other than those of the fishery industry.

Protection of Areas

Another way to protect biodiversity is to protect particular geographic regions, or ecosystem areas, or types of habitat. Several treaties take this approach, many of them including marine ecosystems either as the focus of the agreement or as a part of it.

Polar Regions

The Antarctic continent and the Southern Ocean, which surrounds it, have been the subject of several treaties, first to guarantee common rights to exploit the living and nonliving resources of the area (since it is neither a nation nor the native home of any people) and then to protect them. The Antarctic Treaty was adopted in 1959 to prohibit use of the continent and surrounding waters for military purposes and weapons testing and to promote scientific research. The treaty applies to the Antarctic continent and the ice shelves south of sixty degrees south latitude. The original treaty does not have environmental provisions except to prohibit the disposal of radioactive wastes, but it does invite parties to consider such provisions. The Madrid Protocol was a high-profile negotiation that, after much contention, was signed in 1991. It establishes Antarctica as a natural reserve devoted to peace and science and provides for research and monitoring, waste management, native biota protection, shipping restrictions, specially protected areas, and managed areas. It creates a Committee for Environmental Protection to advise on implementation of the protocol, including the Antarctic Protected Area System, which can include marine areas. Insofar as most Antarctic wildlife spends time in the ocean, this treaty protects marine biodiversity.

However, much of the Southern Ocean, with its rich abundance of life and its biodiversity, is not covered by this treaty. Another treaty, the Convention on the Conservation of Antarctic Marine Living Resources (CCAMLR), was negotiated in 1980 and is still waiting to be ratified. Although this is a fisheries treaty, it is also intended to function as an ecological treaty. Marine resources are defined as including all living marine organisms. The area covered by the treaty has an ecosystem definition rather than a political one and includes all waters south of the Antarctic Convergence, where the cold waters of the Antarctic Ocean meet warm waters from the north. Whales and seals are excluded from the treaty because they are covered by the International Whaling Convention and the Antarctic Seals Convention. CCAMLR sets down several principles of conservation that are to be applied to any fishing and associated activities in the covered area. These include the prevention of a decrease in size of any fished population. Recommended measures include the imposition of fishing levels and methods to ensure sustained populations of the fished species and to maintain ecological relationships between fished species and other species. Any risk of irreversible or long-term (lasting more than two decades) changes in the marine ecosystem is to be avoided or minimized.

The Arctic region, which includes a significant marine environment, is inhabited by people and is divided among several nations. Consequently, it cannot be protected in the same way the Antarctic is. There is no comprehensive environmental treaty for the Arctic, but recently the Arctic nations, including several groups of indigenous peoples, have agreed to cooperate on protection of the environment and on research in the region. The Arctic Environmental Protection Strategy (AEPS) was adopted in 1991 with the objectives of protecting Arctic ecosystems and peoples; providing for sustainable utilization of natural resources; accommodating traditional and cultural values and practices of indigenous peoples; monitoring the state of the environment; and reducing and eliminating pollution. It established several working groups, among them the Arctic Monitoring and Assessment Programme (AMAP) and the Working Group for the Protection of the Arctic Marine Environment. The primary political body to oversee the AEPS is the Arctic Council, with membership of each Arctic nation and indigenous group.

Other Regional Seas

The United Nations Environment Programme (UNEP) has established a large marine conservation program that approaches marine conservation on a regional basis, with regional arrangements to control pollution, manage marine resources, and protect habitats and species. Once a discrete sea or sea area with particular environmental concerns has been designated, UNEP prepares an action plan on which a framework convention is based. The convention is negotiated by the countries bordering that sea area and is supplemented with agreed protocols for implementation of various aspects of the convention. There are currently thirteen regional seas areas, but UNEP support is so weak because of lack of funding that most of the agreements are stronger on paper than in reality and action depends solely on the dedication of participating countries. The program's ambitions are high and encompass a comprehensive approach to environmental problems, especially the need to reduce pollution and to protect natural areas and wildlife. The concept is a good one, focusing on large marine ecosystems and the nations that share them and, therefore, have the most control over them and the most to gain from efforts to maintain their ecological integrity. With increased coverage—all coastal areas of the globe should be covered by regional seas agreements—and increased commitment and action, this could become a most effective approach to marine biodiversity protection.

The UNEP regional seas include the Mediterranean; Black Sea; Kuwait (Persian Gulf); West and Central Africa; East Africa; South Asian Seas; East Asian Seas; coastal Southeast Pacific; Red Sea and Gulf of Aden; the South Pacific; South-west Atlantic; and Caribbean. Four of these have protocols for protected areas and wildlife, and three have protocols for land-based sources of marine pollution. Two other regional seas agreements, negotiated outside UNEP's program, involve the Baltic Sea (the Helsinki Convention) and the North Atlantic (the OSPAR Convention, preceded by the Paris Convention for the North Sea). These have pollution provisions and protocols that have become models for action on marine pollution at the global level.

Protected Areas

Areas and habitats identified as needing protection are usually of such a size that they fall into the jurisdiction of a single country, and it is up to that country to decide whether and how to afford protection. The largest and best-known national marine protected area is the Great Barrier Reef Marine Park in Australia. There is no global legal obligation for nations to protect their coral reefs, but protection has been encouraged by many international organizations, such as UNEP, the IUCN (World Conservation Union), ICLARM (the International Centre for Living Aquatic Resources Management), and the United Nations Educational, Scientific, and Cultural Organization's Intergovernmental Oceanographic Commission (IOC), all of which have programs to heighten awareness of the problem and promote the protection of these habitats. The Global Coral Reef Monitoring Network (GCRMN) is a component of the International Coral Reef Initiative (ICRI), which is sponsored by all these organizations. The network is divided into six regions, within which there is to be collaboration among communities, governments, and scientists to gather data on the condition of reefs globally and to raise awareness among local communities about the problems facing their reefs and potential solutions. The establishment of protected reef areas is encouraged, including fully protected parks, multiple-use sanctuaries, and no-fishing zones.[6]

The United Nations Educational, Scientific, and Cultural Organization (UNESCO) sponsors the Man and the Biosphere Programme, a network of representative protected ecosystems around the world. The idealized model for these reserves includes a core zone, which is strictly protected from human disruption, and a surrounding buffer zone, which can accommodate managed traditional uses that are consistent with the objectives of the reserve. In practice, however, biosphere reserves have not lived up to the model, particularly in coastal and marine areas, where few of these reserves have been designated.[7]

Other marine areas with international protection can be and have been designated within the context of various UNEP programs and other regional seas programs, especially in the Caribbean, the South Pacific, the Mediterranean, East Africa, and Antarctica. Protection from the negative effects of shipping activities is also provided by the International Convention for the Protection of Pollution from Ships (MARPOL 73/78 and the IMO's Marine Environment Protection Committee (MEPC). The committee can designate "special areas," which require special pollution restrictions; "particularly sensitive sea areas," which require special measures to maintain ecological integrity; and "areas to be avoided" by large vessels carrying oil and other hazardous materials.

Finally, the Ramsar Convention on Wetlands of International Importance establishes an international wetlands conservation system based on an official List of Wetlands of International Importance. It promotes the management and conservation of freshwater and marine wetlands by all parties, of which there are approximately eighty. To be a party, a country must have at least one wetland on the list. In addition, parties are required to foster the "wise use" of wetlands, to conserve wetlands and waterfowl, and to establish wetland nature reserves, whether they are on the list or not. Although this convention provides a useful tool for protecting ecological processes and functions along the coasts, particularly through its focus on spawning and breeding areas, there has not yet been an attempt to develop a representative network of marine wetlands.

Marine Area Protection and Management in the United States

The National Marine Sanctuary Program was established in 1972 by the Marine Protection, Research, and Sanctuaries Act (MPRSA). The program's purposes are to identify marine areas of special national significance due to resource or human-use values; to provide for management and regulation of these areas; to promote scientific research and monitoring; to educate the public; and to facilitate multiple use of the designated areas. NOAA is in charge of the program, which oversees the designation, regulation, and management of national marine sanctuaries. There are twelve designated sanctuaries, and two others are under serious consideration, although more than one hundred sites have been nominated. Most sanctuaries have been designated for their biological importance, and those more recently designated are quite large. They include the following:

- *Monitor* National Marine Sanctuary (established 1975), the site of a sunken ship located off the coast of North Carolina, was designated for its historical value.
- Channel Islands National Marine Sanctuary (1980), located off the southern shore of California, protects 3,245 square kilometers (1,252 square miles) of habitat for marine mammals and birds.
- Gray's Reef National Marine Sanctuary (1981), covering a 44-square-kilometer (17-square-mile) area off the coast of Georgia, includes one of the Atlantic coast's largest "live bottom" reefs, consisting of hard and soft corals, near the northern limit of the coral's range, as well as a wide variety of fish.
- Gulf of the Farallones National Marine Sanctuary (1985), located off the northern coast of California, encompasses 2,429 square kilometers (938 square miles) of marine environment surrounding the Farallon Islands. It includes important marine mammal habitat and the largest seabird rookery in the contiguous United States.
- Fagatele Bay National Marine Sanctuary (1985), located in American Samoa, is composed of 60 hectares (160 acres) of coral reef ecosystem.
- Cordell Bank National Marine Sanctuary (1989), a seamount at the edge of the continental shelf about 32 kilometers (20 miles) off Point Reyes, in northern California, provides habitat for a rich variety of benthic and pelagic animals and seabirds.
- Flower Garden Banks National Marine Sanctuary (1990), located in the Gulf of Mexico about 193 kilometers (120 miles) from shore,

includes two banks that are covered with rich coral reef and algal reef communities as well as an unusual ocean brine seep community associated with an area of high salinity seepage from an underground salt deposit.

- Florida Keys National Marine Sanctuary (1990) covers 6,734 square kilometers (2,600 square miles) of coral reefs and surrounding marine environment on both the Straits of Florida and Florida Bay sides of the Florida Keys.
- Hawaiian Islands Humpback Whale National Marine Sanctuary (1992), incorporating 3,367 square kilometers (1,300 square miles) of shallow warm waters surrounding the main Hawaiian Islands, is one of the world's most important humpback whale habitats—the winter breeding ground for an estimated two-thirds of the total population of this endangered species.
- Monterey Bay National Marine Sanctuary (1992), spanning more than 13,727 square kilometers (5,300 square miles) of coastal waters along the central coast of California, incorporates a rich array of habitats, from rocky shores and lush kelp forests to one of the deepest underwater canyons on the West Coast. It is one of the richest areas of marine biodiversity in the coastal United States.
- Stellwagen Bank National Marine Sanctuary (1992), 2,181 square kilometers (842 square miles) of ocean and shoals lying at the mouth of Massachusetts Bay and enriched by an upwelling of nutrient-rich water from the Gulf of Maine, supports high productivity and a diverse food web, including several species of great whales, and provides a refuge within the overfished area off the Atlantic coast.
- Olympic Coast National Marine Sanctuary (1994), encompassing 8,573 square kilometers (3,310 square miles) and extending nearly 64 kilometers (40 miles) outward from Washington's outer coastline, provides habitat for one of the most diverse marine mammal faunas in North America and is a critical link in the Pacific flyway. It supports a rich mix of Native American culture as well.

The national marine sanctuaries offer both an opportunity and a challenge for effective integrated coastal zone management. Each must be regulated and managed with the cooperation of the state whose coastal waters encompass or are part of the sanctuary. Thus, some of the most important threats to a sanctuary are also those that are the most difficult to get the state to agree to regulate strictly. For instance, the Florida Keys National Marine Sanctuary is threatened most by coastal residential development and by sport fishing—the two things most precious to people onshore. The challenge is to make residents appreciate the fact that stricter regulation of septic and other drainage and limitations on fishing and development will be to their benefit and that of their beloved patch of ocean. The other challenge is to make the sanctuaries effective with the low levels of funding approved by Congress.

Another U.S. coastal zone management initiative is the Coastal Zone Management Act (CZMA). Enacted in 1972 and most recently reauthorized in 1995, the CZMA provides for a system of incentives for states that establish certain land-use controls along their coasts. Coastal states are required to identify particular areas of concern, including exceptional, rare, fragile, or vulnerable natural habitats; areas of high natural productivity; essential habitats for wild species; coastal protection areas, particularly dunes, reefs, beaches, sandbanks, and mangroves; and zones necessary for the maintenance or supply of coastal resources, such as coastal floodplains, and zones essential to recharging of the water table. The areas in question are to be protected through land-use regulations adopted by state or local authorities in accordance with state regulations as well as federal requirements. Most of the coastal states have availed themselves of this opportunity and have developed their own coastal zone programs approved by the federal government. The CZMA also provides for the conservation of remaining natural estuaries through grants to the states for acquiring land necessary to accomplish this. Once secured, the estuaries and adjacent associated habitats must be managed as research natural areas.

Two other acts provide management regimes for marine areas. The Coastal Barrier Resources Act (CBRA) establishes a Coastal Barrier Resources System composed of thousands of miles of barrier islands along the Atlantic and Gulf coasts, where federal money cannot be spent on infrastructure for development of the islands. This act is indirectly protective of shorebirds, sea turtles, and estuaries. The Outer Continental Shelf Lands Act encourages exploration and development of hard minerals, oil, and gas on the outer continental shelf but stipulates that such development should occur within environmentally acceptable guidelines. Resource development generally outweighs environmental protection in the application of this act.

Marine Pollution

Control and elimination of marine pollution are critical in protecting marine biodiversity, and as has been seen, they are an important element in the management of protected areas, in regional seas programs, and in integrated coastal zone management. They are also the subject of several international treaties and U.S. laws.

Pollution Caused by Marine Activities

The Convention on the Prevention of Marine Pollution by Dumping of Wastes and Other Matter—first known as the London Dumping Convention, or LDC, and subsequently known as the London Convention, 1972, or LC72—deals with marine disposal of wastes generated on land. Initially, it permitted but regulated dumping of industrial wastes, sewage sludge, and dredged materials; incineration of wastes on vessels or barges at sea; and transport to the sea of any other wastes or materials for the purpose of dumping. The treaty prohibits marine dumping of wastes containing significant amounts of certain substances on a "black list." Wastes containing other substances that are listed on a "grey list" can be dumped only with a special permit, which is granted if the responsible government agency has determined that the disposal would not present a threat to the marine environment. Any wastes containing no listed substances require a general permit. Over the years, the contracting parties have prohibited one after another of the categories of wastes typically dumped at sea—low-level radioactive wastes (high-level radioactive substances were on the black list to begin with), wastes to be incinerated at sea, and industrial wastes. The only major categories of potentially polluting wastes still allowed within the treaty's criteria are dredged materials and sewage sludge. The change of name (dropping of the word *Dumping*) came about as the treaty became more restrictive and parties became sensitive to its early reputation as a "dumpers' club." Dumping accounts for less than 20 percent of marine pollution, but the effectiveness of this treaty has been an important contribution to its reduction. The treaty's success has been attributed in part to its association with the effective International Maritime Organization, which regulates international shipping. Although it is an independent treaty, it uses the IMO secretariat. Certain NGOs, especially Greenpeace International, have been instrumental in the progress of this treaty.[8]

Another treaty dealing with routine discharges associated with normal shipping operations is an IMO treaty. The International Convention for the Prevention of Pollution from Ships, better known as MARPOL or MARPOL 73/78, covers accidental spillage or purposeful dumping of ship cargo or wastes generated on board. The treaty has five annexes, each dealing with separate pollution sources: (I) oil pollution, (II) noxious liquid substances in bulk, (III) harmful substances in packaged form, (IV) sewage, and (V) nondegradable plastics and garbage. Discharge of oily wastes, sewage, and garbage is restricted, and special precautions must be taken to prevent the loss of potentially harmful cargo. Dumping of plastics is prohibited. The Marine Environment Protection Committee (MEPC) of the IMO is responsible for advising on MARPOL issues. The IMO also deals with requirements generated by this treaty for waste reception facilities in ports. Several environmental NGOs have been active in meetings and committees of MARPOL—Friends of the Earth, World Wide Fund for Nature, and Greenpeace in particular.

U.S. laws that deal with pollution from ships include the Ocean Dumping Act (part of the Marine Protection Research and Sanctuaries Act, or MPRSA); the Ocean Dumping Ban Act; the Marine Plastics Pollution Research and Control Act; and the Oil Pollution Act of 1990 (OPA 90). The Ocean Dumping Act controls marine dumping of all materials, including sewage sludge, industrial waste, and dredged materials, and the subsequent Ocean Dumping Ban Act amends that act by banning marine disposal of industrial wastes and sewage sludge. The EPA and the U.S. Army Corps of Engineers are responsible for developing the regulations that implement the law and for issuing dumping permits. Specially designated sites for dumping materials (now restricted to dredged materials) are required, and dumping is not supposed to result in degradation of the marine environment outside the dump site. The Marine Plastics Pollution Research and Control Act bans marine disposal of plastics, including synthetic fishing nets, within the U.S. EEZ and by U.S. vessels anywhere, and it implements that provision of MARPOL.

The Oil Pollution Act of 1990 was passed in the wake of the infamous *Exxon Valdez* oil spill in Prince William Sound, Alaska. It defines stiff terms of liability for oil spills and is designed to provide for rapid response in cleanup and damage assessment and rapid payment of damages to those incurring loss or injury from an oil spill. The objective of the act is to provide a financial deterrent that will foster stronger measures on the part of oil companies to prevent future spills.

Several United Nations organizations established the Joint Group of Experts on the Scientific Aspects of Marine Pollution, known as GESAMP (the word *Pollution* was subsequently changed to *Environmental Protection,* but the acronym remains the same). This is a body of experts, primarily scientists, who advise members of United Nations organizations and parties to treaties about particular issues and questions related to marine pollution. They organize working groups within subcommittees and produce reports on their deliberations.[9]

Land-Based Sources of Marine Pollution

Activities on land are responsible for about 80 percent of marine pollution. There are regional seas treaties that address land-based sources, but there is no comprehensive global treaty addressing the problem. However, the Global Programme of Action for the Protection of the Marine Environment from Land-Based Activities, mentioned earlier, recommends numerous actions that can be taken at national, regional, and international levels to reduce pollution from land. Among the most important international actions is the commitment to the negotiation of a treaty addressing the manufacture and use of persistent organic pollutants (POPs), with a suggestion for an initial ban on twelve of the most common, persistent, and destructive of these chemicals. Negotiation of this treaty began in 1998, with an ambitious goal of adoption by 2000.

Two existing treaties offer approaches to the problem of global movement of persistent toxic substances. The 1989 Basel Convention on the Control of Transboundary Movements of Hazardous Wastes and Their Disposal bans the export of toxic waste from any country belonging to the Organization for Economic Cooperation and Development (OECD) to any non-OECD country (i.e., from an industrial country to a developing country) for the purpose of disposal or recycling. The 1979 Convention on Long-Range Transboundary Air Pollution (LRTAP), adopted by northern industrial countries, addresses the reduction of toxic airborne emissions, with separate protocols on sulfur emissions, volatile organic compounds, nitrous oxides, and POPs.

In the United States, the major law addressing water pollution is the Clean Water Act, enacted in 1972 for the purpose of maintaining and restoring the chemical, physical, and biological integrity of U.S. waters. To accomplish this, Congress established a combined federal and state system of controls to implement clean water programs. The act focuses on sewage treatment plants and pollution control programs, with regulatory requirements for industrial and municipal discharges.

The Environmental Protection Agency (EPA) has primary authority for developing regulations, setting standards, and issuing discharge permits. The agency's permit system, called the National Pollutant Discharge Elimination System (NPDES), has not eliminated discharges, but it has reduced their levels in many cases. The act provides protection for the marine environment by controlling discharges into all waterways and by regulating direct discharges into marine environments with the objective that they must not "unreasonably degrade the marine environment" within the territorial sea. Criteria for determining degradation include adverse changes in diversity, productivity, and stability of the biological community; threats to human health; and loss of aesthetic, recreational, scientific, or economic values.

Estuaries are not effectively protected by these criteria, but the act set up the National Estuary Program to address estuarine and coastal pollution by developing management plans for specific estuaries. Environmentalists believe the Clean Water Act could be improved by adding provisions for better protection of wetlands and increased prevention and cleanup of toxic water pollution.

Finance and Trade

Although a detailed examination of economic institutions is beyond the scope of this book, it should at least be mentioned that their policies can have a significant effect on the environment in general and on marine biodiversity in particular. Multilateral development banks (such as the World Bank, the Inter-American Development Bank, and the Asian Development

Bank) are institutions through which money from rich countries of the developed world is funneled to poor countries of the developing world through loans granted for specific projects. All countries participating in a given transaction are members of the involved institution, and governments of the countries providing the money have representatives on a board that approves or disapproves loans. In theory, the borrowing country conceives a project and submits a proposal, but in practice, the countries providing the money tell the borrowing countries what types of projects will be funded.

Through their roles in providing money for development projects in developing countries, multilateral development banks and bilateral lending institutions, such as the U.S. Agency for International Development (AID), have caused a great deal of ecosystem degradation in developing countries to which money was loaned or granted. Certain kinds of projects they promote, such as dam building and industrial development, sometimes in coastal areas, often have had serious environmental consequences. In response to public outcry led by several environmental organizations denouncing such projects, these institutions have established environmental departments and have developed environmental criteria for their projects. The rhetoric is still much more protective than is the practice, however, and it is still important for parties interested in changing the system to monitor and protest projects that threaten the environment.

One general policy of these economic institutions that tends to create problems is a preference for large-scale, high-tech projects, such as construction of very expensive dams and sewage treatment plants. Sometimes these create more environmental problems than they solve, when smaller projects would be better. In the case of sewage treatment, for example, constructed wetlands or, in some cases, modern biological toilets would provide more affordable and more effective ways of managing human organic wastes. Furthermore, the presence of sewage treatment plants tends to encourage disposal of industrial wastes into the water-based sewage collection system, a problem that could be avoided with smaller local solutions.

Another policy that has been destructive is the common emphasis on stimulating economic development and international trade rather than on improving the personal living conditions of the majority of citizens. Marine biodiversity, for instance, is threatened by these economic institutions' persistent funding of environmentally unsound coastal aquaculture development projects aimed at feeding international markets rather than hungry people within the country.

A different international funding mechanism, focusing solely on environmentally beneficial projects in developing countries, is the Global Environment Facility (GEF), which was created in 1991 with implementation jointly in the hands of the United Nations Development Programme (UNDP), UNEP, and the World Bank. It provides financing, usually in the form of grants rather than loans, for projects that will benefit the global environment in four areas, called "focal areas." These are biodiversity, climate change, ozone depletion, and international waters. The GEF is the primary funding source for the Convention on Biological Diversity. Any government may become a participant, which means being allowed a role in decision making and governance. There is an expectation that richer countries will commit money. The biodiversity and international waters focal areas are those most important to the conservation of marine biodiversity. High-priority problems to be addressed in projects within international waters include control of land-based sources of water pollution; prevention and control of land degradation that affects international waters; prevention of physical and ecological degradation; control of unsustainable exploitation of living resources; and control of ship-based sources of chemical pollution and species introductions.

Trade treaties can have serious implications for natural ecosystems by undermining national laws that protect species and ecosystems from negative effects of commercial exploitation. In the context of two major trade treaties, the General Agreement on Tariffs and Trade (GATT) and the North American Free Trade Agreement (NAFTA), there have been challenges to national laws prohibiting importation of certain products that have been captured or produced in a way that threatens species or ecosystems. For example, tuna caught by fishing methods that endanger dolphins cannot legally be imported into the United States, but GATT ruled this to be an unfair trade practice. The United States has taken a position in favor of building protection of national environmental laws into trade agreements.[10]

The Role of Nongovernmental Organizations

Persistent pressure by grassroots, national, and international environmental groups calling for programs, laws, and policies that favor sustainable development and conservation of biodiversity play an important role in precipitating changes in national and international policies and actions. Timothy Wirth, undersecretary of state for global affairs in the U.S. Department of State, described the role of nongovernmental organizations (NGOs) in the quest for "a sustainable equitable balance between the demands of humanity and the planet's capacity to support life": "The heroes and heroines of the Rio conference were not governments but the non-governmental organizations around the world that defined the agenda for Rio and the Earth Summit's Agenda 21 document. It was out of their work that this remarkable document came."[11]

The principal functions of these groups are to (1) educate the public and government officials; (2) analyze current policies; (3) advocate on behalf of citizens to achieve effective conservation policies, legislation, and international agreements; and (4) help oversee the implementation of protective laws and agreements. In the case of marine biodiversity, the educational role of involved NGOs is critical as laws are implemented and relevant treaties are negotiated.

Some of these groups have provided valuable analyses of government policies on biodiversity, fisheries and aquaculture, ocean pollution, and coastal development. At the international level, some countries rely heavily on the most responsible of these assessments. Environmental NGOs actively promote particular changes in policies and legal regimes regarding the environment in general and marine biodiversity in particular, and in doing so they often provide support for governments to move further toward conservation and sustainable policies than they otherwise would have.

Rarely, however, are the goals of the NGOs fully achieved. Once good policies and legal regimes are in place, the NGO oversight role is critical because there is often a tendency for agencies and officials responsible for implementation to be slow or deliberately weak in accomplishing the prescribed changes. Grassroots groups are as important to oversight of marine environmental initiatives as are national and international organizations because they are located at the site of implementation—on the coasts, where serious degradation of marine ecosystems must be addressed.

Among the largest international environmental organizations are the World Wide Fund for Nature or the World Wildlife Fund, Greenpeace International, and Friends of the Earth International, all of which routinely send delegations of observers to meetings of relevant treaties and international institutions. An organization that falls somewhere between government and nongovernmental is the World Conservation Union (also still known by its former acronym, IUCN). This is an independent organization with both government and nongovernmental members, and as such it is moderate in its policies and influential in international affairs. IUCN promotes conservation of biodiversity through policy statements, publications, and participation in international forums. In formal intergovernmental negotiations, it has an observer role like that of the NGOs.

Some reviewers of the first edition of this book thought that its nod to NGOs was self-serving. However, NGOs play an undeniably crucial role in linking citizens with their government officials and in carrying citizens' concerns into the ethereal realm of international negotiations.

Chapter Eight

Can Marine Biodiversity Be Saved?

Despite all the efforts to address different aspects of the threat to biodiversity on land and in the ocean, there is still no broad ethic that promises to change human behavior so that we finally cease to pose a huge threat to the survival of the natural world as we know it. Until such a sweeping change comes about, well-meaning efforts to stop the biological hemorrhaging can only temporarily bandage the old wounds as we continue to inflict new ones. We have waged war on our most supportive allies—our own ecosystems on land and the living ocean. We must make peace before we destroy them altogether and find ourselves alone without a world to support us.

It is not an easy task and it requires both leadership and guiding ethical principles. We are beginning to see leaders emerging from the fields of science, government, business, media, religion, and others concerned. Hopefully their words will guide new ethics and actions by people and their governments around the world. There are also guiding principles that have been acknowledged internationally and now need to be put into practice. While the focus of this book is marine biodiversity, the principles apply to all life on earth and nothing specific can save ocean species until a sweeping change in human attitudes is realized. Thus the rest of this chapter deals with cries of concern that are increasingly being heard and glimmers of hope that such a change may occur.

Warnings

Scientists are beginning to grasp the enormity of the losses in biodiversity taking place worldwide, and are speaking out about their concern at public forums and in the prolific scientific literature on the topic. "Current global environmental changes that affect species composition and diversity are . . . profoundly altering the functioning of the biosphere," stated a 1998 paper by several ecologists; and the respected British ecologist Norman Myers asserted in 1993, "Whether in biological, ecological, environmental or economic terms, mass extinction will clearly constitute an irreversible impoverishment of the Earth and of the world. That much is certain; the rest is uncertain."[1]

Those in the environmental field are constantly admonished not to be so pessimistic and to have an upbeat message for the public because they will surely not listen if the message is gloomy. Consequently, there are numerous lists of ten-things-you-can-do to save the planet. As well meaning as these lists are and as important as it is for individuals to take responsibility for their own actions, in truth, far more than that is now needed. Certainly there is hope, but for that hope to become reality, the actions of individuals and those of governments must be based on a truthful assessment of the gravity of the situation, not on the fiction that ten simple steps will avoid disaster. New strong leadership is needed.

The loss of biodiversity is often viewed by the public and policy makers as being much the same as issues like pollution, nuclear proliferation, and problems of international economies—as inevitable ills that can be fixed in due course. Yet, although often those other problems can indeed be resolved within a few generations or less if the will is there to do so, the loss of biodiversity is final. Yet, there is not a great deal of concern on the part of the public and government. Why? Because, notwithstanding the losses of biodiversity that have already occurred, we in the developed—some might say overdeveloped—world have noticed few direct repercussions in the way we live our lives. Indeed, one scientist has expressed the fear, shared by others, that the earth may suffer tremendous losses of biodiversity and impoverishment of natural ecosystems while humanity, aided by technological adaptations, survives for some time in a state of false security.[2]

Furthermore, decision makers in governments tend to pay attention only to the consequences of biodiversity loss that directly and immediately affect the lives of people, primarily those people who are influential enough to be part of the political process. Among the more affluent, these are the very people most likely to be sheltered from nature by technology. Conversely, the people who rely most directly upon the natural world—many of the indigenous peoples, small-scale fishers, and the like—are the first to be affected, to know they are being affected, and to say so, but the last to be heard.

At some point, however, it is certain that the loss of biodiversity will be acutely felt by all, as the failure of biologically impoverished ecosystems to adapt to further changes in the global environment results in the loss of one life-support system after another and the inability of the natural world to meet the needs of humans and other species that remain. As E. O. Wilson—the Harvard scientist whose name has become synonymous with biodiversity because of his extensive writing and speaking on the subject—says in his masterpiece of insight, *Biophilia,* "This is the folly our descendants are least likely to forgive us."[3] Not only will humans suffer the consequences of failing ecosystems, but also—even if we do survive for some time—as other species disappear, the innate feeling of kinship humans have with other life-forms will increasingly be stifled.

Alarms are being raised by biologists who study biodiversity in natural ecosystems. They are witnessing the impoverishment of numerous biological communities; but politicians and regulators often will not listen to them unless economic or human health arguments are offered as well. Thus, some scientists have joined with economists in the exercise of assigning monetary values to nature's "goods and services." Being able, at last, to ascribe quantitative human-relevant economic values to living natural systems is intoxicating for those who have tried for so long with so little success to get the attention of the public and of government and to warn them of the importance of maintaining the integrity of natural ecosystems and conserving living species in their natural habitats. By assigning monetary values it has been possible at last to illustrate some of the "intimate connections" described by noted marine biologist Jane Lubchenco, "between these systems and human health, the economy, social justice, and national security."[4]

Perhaps, however, it is a mistake to base decisions entirely or primarily upon quantitative economic and scientific factors without including ethical and spiritual values. E. O. Wilson has described a natural progression from one to the other:

> When very little is known about an important subject, the questions people raise are almost invariably ethical. Then as knowledge grows, they become more concerned with information and amoral, in other words more narrowly intellectual. Finally, as understanding becomes sufficiently complete, the questions turn ethical again. Environmentalism is now passing from the first to the second phase, and there is reason to hope that it will proceed directly on to the third.[5]

At the end of the twentieth century, we are firmly planted in the second (amoral) stage, guided by an affinity for scientific data and technological solutions and a tendency to place humans outside and in control of the natural world. Applying economics and monetary valuation to nature is merely an extension of this. Such an exercise prolongs our tenure in Wilson's amoral phase, as do the worship of science, risk assessment, capacity estimations, and cost-benefit analyses.

Canadian ecologist and communications expert David Suzuki has pointed out that the economy should be a subset of the environment, but people think of it the other way around—seeing the economy, not the natural world, as the provider of everything.[6] Suzuki has further expressed frustration and disappointment over the fact that, despite an undeniable interest in nature, the public has little inclination to act on its behalf; and Peter Raven, an expert on plant biodiversity, has noted our tendency to get bogged down in legalistic details as we dedicate ourselves to conservation.[7]

Something more is needed to assist our progression from a decision-making process dominated by science into one that incorporates ethical values, wherein science informs but does not decide. There are a few glimmers of hope that we may be proceeding, though perhaps more slowly than Wilson prescribes, into the ethical phase of environmentalism.

Guiding Principles

The route may be best charted by applying the precautionary principle, which has been nominally accepted by the international governmental and nongovernmental establishment. This principle recognizes that even science is filled with uncertainty and bias, and while emphasizing the importance of science in providing information, it ultimately encourages the final decisions to be based on moral judgment. As described in earlier chapters, this concept has been incorporated in the language of numerous international environmental agreements—Agenda 21; the Convention on Biological Diversity; the London Convention, 1972; and the 1995 high seas fisheries agreement added to UNCLOS. Yet it has never really been implemented in any meaningful way. We have numerous international treaties and national laws that, if implemented with the guidance of the precautionary principle, would provide improved protection for the natural world and human welfare. Furthermore, if individually we applied it to our daily lives, we would have a better sense of our own role in nature and our power to conserve it. If fully implemented on all levels—from international treaties to personal actions—the precautionary principle could serve as something of a golden rule of the environment.

The first element of the precautionary principle is to gather available scientific information, including traditional knowledge from indigenous cultures, and to recognize the uncertainties and missing information. The knowledge is then used to estimate the potential effects of our proposed activities, allowing for surprises. Finally appropriate action is taken to prevent significant harm to ecosystems even in the absence of scientific certainty. Lack of scientific proof of anticipated effects is not a reason for taking no action or postponing action to prevent or minimize potential harmful effects.[8]

The fundamental premise of the precautionary principle is to *prevent* harm to ecosystems and to humans. The principle should be implemented through environmentally sound alternatives that still can accomplish the goals of potentially harmful activities, *if* those goals are justified. Economics do not govern precautionary decisions, but obviously the implementation of a decision may be affected by monetary concerns. Consequently, those with the money and/or technology to implement precautionary actions should aid those less able to carry them out. In fact, this has become a major issue between "developed" and "developing" countries in many international negotiations.

A moral mandate associated with the precautionary principle, and also with sustainable development, is an obligation to provide for the needs and rights of future generations as well as the present. E. O. Wilson's warning that future generations will not likely forgive us for extinguishing species should move us to act. As Norman Myers has so aptly put it:

> Right now, we are effectively asserting that we can *afford* to allow large numbers of species to become extinct on the grounds that we cannot economically deploy the funds and other conservation resources necessary to save a good share of vulnerable species. The corollary of this stance is that we are implicitly deciding that at least 200,000 future generations can certainly do without large numbers of species, and that we feel sufficiently certain we know what we are talking about when we make that decision on their unconsulted behalf.[9]

It is time to make the decision in a more deliberate, moral, and precautionary way. We cannot expect to prevent all anthropogenic extinctions, given the major influence our mere presence has upon life on earth. However, we should try to prevent all those we can, recognizing that we will sometimes fail.

Marine ecosystems are in as urgent need of such actions as are terrestrial ecosystems. The precautionary principle applies both directly and indirectly to marine biological diversity. Guided by this principle, individuals, industries, governments, and international institutions may separately and collectively decide to reduce nutrient pollution and work toward the elimination of toxic pollution and persistent debris; to fish in a way that maintains fully functioning populations of target species and eliminates bycatch; and to use and live in coastal marine ecosystems in a way that maintains their natural integrity.

Leaders

The precautionary principle changes the basic questions we ask about our interactions with the environment. Instead of asking how much we can get away with before the damage becomes intolerable, we should ask whether there is some other way to accomplish a task that will prevent, or at least reduce, damage long *before* we reach the stage of unacceptable environmental degradation. Precautionary action calls for innovation—a return to the age of invention, but invention informed by the knowledge that processes and products must be compatible with the living world. The innovative process must be one that does not rely solely on science and one in which we all can participate. Norman Myers has challenged us to move away from a focus on mobilizing the planet's resources to benefit the human cause and toward a focus on mobilizing ourselves to benefit the cause of the planet.[10] To do this will require strong leadership—from Myers and other like him, who understand the severity of the problems and are innovative in devising possible solutions. Other likely leaders are becoming more vocal and visible and with luck they will increasingly lure followers. Some have already been quoted in this chapter and others are mentioned ahead, and they are exemplary of men and women from many different countries, institutions, and disciplines who demonstrate wisdom and insight that give reason to hope we may succeed in shifting the focus from the monetary wealth of people to the natural wealth of the planet.

Paul Hawken, a national leader in environmentally sensitive business practices, has suggested that the next "revolution" will be a new type of industrial revolution based on the flow of services instead of increased production. In this move from goods to services, industries will become responsible for their products, from the acquisition of raw materials through the manufacture, sale, and use of the product until the end of its useful lifetime and beyond to reuse of the materials for another product—thereby reducing the quantity of materials flowing through the system.[11] Thus the products are not only the goods but also the services that go with their production and recycling or reuse. A similar view has been stated by another innovative American businessman, Paul Bierman-Lytle, who envisions "industry as embedded in interrelated systems so that the output [the waste] of one system becomes the input for another."[12] Such a philosophy can also be applied to services needed in the context of communities or urban development. John Todd and Nancy Jack Todd have worked for three decades designing constructed ecosystems to purify water, produce food, and restore environments, both natural and urban, to a healthy state.

They attempt to solve big environmental problems with small-scale solutions that communities of people can implement in a way that is meaningful to their lives. Among their innovations is the cleansing of waste water by "living machines," which they describe as follows:

> Based on the precepts that waste is a resource out of place and that nature handles every form of waste by turning it into a resource, our systems imitate the purifying and recycling abilities of natural aquatic ecosystems. They contain populations of bacteria, algae, microscopic animals, snails, fish, flowers, and trees. These living machines are capable of advanced water treatment without resorting to the hazardous chemicals used in most existing treatment plants and at less than traditional secondary treatment costs.[13]

Innovations such as those described above indeed point to a new industrial revolution inspired by a precautionary approach based on clean production.

New approaches to conservation are also needed—approaches that are more precautionary and less reactive. Jerry Schubel, marine ecologist and director of the New England Aquarium, has suggested "inverse triage," whereby we first protect those habitats that are in good condition, then tend to those most likely to benefit from help, and only then apply heroic measures to the hopeless cases. Our current approach moves in the opposite direction, protecting the most severely endangered habitats and species first. Schubel's approach makes sense, since protecting unendangered, fully functioning ecosystems, species, and populations will prevent them from becoming dysfunctional and should save more species and ecosystems in the end.[14]

Tim Wirth was an extraordinary environmental leader and visionary in the United States government before joining forces with the media leader Ted Turner to select United Nations environment and social programs to be recipients of large sums of money that Mr. Turner has decided to contribute. In Wirth's words, "the world is desperately in need of a new set of shared global values—common purposes grounded in ethical principles of justice and stewardship."[15] To achieve such a switch in focus, ethical values will become as important a part of development as economic values now are. This imperative can be put no better way than is stated in the preface to the proceedings of a 1995 World Bank conference on ethics and spiritual values in promoting sustainable development:

> Two realizations quickly became clear: that values lie at the very heart of our behavior and that sustainable devel-

> opment will occur only when we have belief systems that respect all life, assign priority to the common good, engender responsibility for the whole, promote equality, and support unconditional caring.[16]

Perhaps the precautionary principle can provide a pathway for governments, organizations, industry, and citizens trying to reach this lofty goal—a bridge linking science and economics with ethics and spiritual values.

The world's ocean may offer excellent opportunities to begin applying the precautionary principle in earnest, since it is still possible to prevent damage to the majority of marine ecosystems despite the severely degraded condition of some coastal systems. It is not widely recognized how dependent we have become upon marine biodiversity and the supportive functions of marine ecosystems as we have displaced one functioning terrestrial ecosystem after another. It is not coincidental that a 1997 symposium attempting to unite the goals of religion, science, and environmental protection into a common vision was held on a ship in the Black Sea, one of the most threatened marine ecosystems in the world. In closing the proceedings, Sylvia Earle, one of the world's courageous marine biologists and divers and an environmental leader in her own right, made a statement that provides, as well, a fitting conclusion to this book:

> Combining the knowledge and wisdom of science with the sensitivity of diverse religions [will] create a new and effective ethic for caring for nature starting with the greatly stressed Black Sea environment. In fact, the thoughtful responses of the unlikely, but congenial mix of Patriarchs, Holinesses, Highnesses, Excellencies, scientists, economists, policy-makers, businessmen, press and others during the days of deliberation, while sailing over these troubled waters, suggests that the process is already well underway.[17]

Notes

All references cited here by author and date are listed with complete information in the bibliography.

Foreword

1. E. O. Wilson, 1993.
2. P. H. Raven, 1995.
3. M. L. Reaka-Kudla, 1997.

Introduction

1. R. C. Lewontin, 1990.

Chapter One

1. J. F. Grassle, 1989; J. Gage and P. Tyler, 1991; N. Coleman, A. Gason, and G. Poore, 1997.
2. J. Carlton and J. Geller, 1993; S. J. Gould, 1991.
3. U.S. Congress, Office of Technology Assessment, 1987a. (The Office of Technology Assessment was abolished by Congress in 1997.); G. C. Ray, 1988.
4. G. C. Ray, 1988.
5. J. H. Steele, 1985.
6. F. S. Chapin III et al., 1997.
7. George Woodwell, speaking at the National Academy of Sciences conference "Nature and Human Society: The Quest for a Sustainable World," October 27–30, 1997, Washington, D.C.; J. Cairns Jr. and J. R. Pratt, 1990.
8. M. Holdgate, 1990.
9. J. E. Lovelock, 1979.
10. R. J. Charlson et al., 1987; D. Lindley, 1988.
11. J. Sarmiento, J. Toggweiler, and R. Najjar, 1988; J. R. Toggweiler, 1988.
12. B. W. Frost, 1996; K. Banse, 1990; W. S. Broecker and G. H. Denton, 1990; J. L. Sarmiento, 1991.
13. G. Bigg, 1996.
14. J. Gribbin, 1988.
15. Population Action International, Population and Environment Program, poster titled "Catching the Limit: Population and the Decline of Fisheries" (Washington, D.C.: Population Action International, 1995).
16. World Resources Institute, 1987.
17. M. E. Hay and W. Fenical, 1996; W. Fenical, 1996.

18. N. Myers, 1979.
19. R. R. Colwell, 1983, p. 19.
20. M. Fox, 1996.
21. H. A. Mooney et al., 1995; R. Costanza et al., 1997.

Chapter Two

1. N. Myers, 1990; J. Cairns Jr., 1987; R. C. Lewontin, 1990.
2. N. Myers, 1990; G. J. Vermeij, 1991; R. Barbault and S. Sastrapradja, 1995.
3. World Resources Institute, 1987; C. Safina, 1995.
4. L. W. Botsford, J. C. Castilla, and C. H. Peterson, 1997; S. A. Earle, 1995; J. Lubchenco, 1998.
5. S. A. Earle, 1995; P. Weber, 1993.
6. D. Pauly et al., 1998; L. W. Botsford, J. C. Castilla, and C. H. Peterson, 1997.
7. L.W. Botsford, J. C. Castilla, and C. H. Peterson, 1997; P. M. Vitousek et al., 1997; P. K. Dayton, 1998; National Research Council, 1995.
8. P. K. Dayton, 1998; S. C. Jameson, J. W. McManus, and M. D. Spalding, 1995.
9. J. Ryther et al., 1972; Pacific Congress on Marine Science and Technology, 1995.
10. R. Tiner, 1984; M. D. Fortes, 1988; National Research Council, 1995.
11. National Research Council, 1995; M. Rozengurt and I. Haydock, 1993.
12. A. M. Manville, 1988.
13. GESAMP, 1990; J. J. Stegeman, P. J. Kloepper-Sams, and J. W. Farington, 1986; AMAP, 1997; D. Livingstone, P. Donkin, and C. Walker, 1992.
14. A. Conversi and J. McGowan, 1994; GESAMP, 1990; G. Mayer, 1982; UNEP, 1995.
15. P. M. Vitousek et al., 1997.
16. J. Hardy, 1991.
17. K. Schmidt, 1997; S. A. Earle, 1995.
18. GESAMP, 1990; EPA, 1997.
19. GESAMP, 1990.
20. D. E. Kime, 1995; W. J. Langston, N. D. Pope, and G. R. Burt, 1992.
21. B. S. Shane, 1994.
22. T. Colborn, D. Dumanoski, and J. P. Myers, 1997.
23. G. J. Vermeij, 1991.
24. J. Carlton and J. Geller, 1993.
25. J. Hedgpeth, 1993.
26. D. Roemmich and J. McGowan, 1995; J. A. McGowan, D. B. Chelton, and A. Conversi, 1996.
27. D. Karentz, 1992.

Chapter Three

1. S. L. Pimm, 1984.
2. A. S. Moffat, 1996; H. Gitay, J. Wilson, and W. Lee, 1996; P. L. Angermeier and J. R. Karr, 1994.
3. E. R. Pianka, 1988, pp. 309–314.
4. B. A. Menge, 1992; UNEP, 1995; D. Raffaelli and S. Hawkins, 1996; P. K. Dayton, 1992.
5. B. A. Menge, 1992.
6. P. K. Dayton et al., 1988.
7. P. K. Dayton, 1998, p. 821.
8. R. S. Burton, 1983; J. C. Gallagher, 1980.

9. National Research Council, 1995.
10. U. Gyllensten and N. Ryman, 1985; P. Klerks and J. S. Levinton, 1989.
11. P. J. Smith and Y. Fugio, 1982.
12. J. P. Grassle and J. F. Grassle, 1976; National Research Council, 1995.
13. National Research Council, 1995; D. Malakoff, 1997; M. L. Reaka-Kudla, 1997.
14. J. Gage and P. Tyler, 1991.
15. L. R. Pomeroy, 1992; S. W. Chisholm, 1992; H. A. Mooney et al., 1995.
16. National Research Council, 1995.
17. G. C. Ray, 1988; R. M. May, 1988.
18. J. Briggs, 1994; R. M. May, 1994a; M. L. Reaka-Kudla, 1997; J. F. Grassle and N. J. Maciolek, 1992; G. C. B. Poore and G. D. F. Wilson, 1993.
19. G. C. Ray, 1988; National Research Council, 1995; R. M. May, 1994a; M. L. Reaka-Kudla, 1997.
20. A. R. Emery, 1978.
21. National Research Council, 1995.
22. H. L. Sanders, 1968.
23. E. C. Pielou, 1979.
24. D. E. Morse and A. N. C. Morse, 1988; G. D. Ruggieri, 1976; R. R. Colwell, 1983; J. H. S. Blaxter and C. C. Ten Hallers-Tjabbes, 1992.
25. D. L. Hawksworth and M. T. Kalin-Arroyo, 1995; M. A. Rex et al., 1993; F. G. Stehli, R. G. Douglas, and N. D. Newell, 1969; M. Angel, 1993.
26. M. Angel, 1993.
27. National Research Council, 1995.

Chapter Four

1. J. Hedgpeth, 1993; D. S. McLusky, 1981.
2. D. F. Boesch, 1974.
3. L. Deegan, 1993; G. C. Ray, 1997.
4. R. G. Wiegert and L. R. Pomeroy, 1981; F. Short and S. Wyllie-Echeverria, 1996.
5. S. Nixon, 1997.
6. National Research Council, 1995.
7. National Research Council, 1993; Y. P. Zaitsev, 1992.
8. E. Ricketts and J. Calvin, 1962.
9. S. A. Earle, 1991; P. J. Bryant, 1998.
10. E. J. Farnsworth and A. M. Ellison, 1997; R. Ricklefs and R. Latham, 1993; R. Bossi and G. Cintron, 1990.
11. EPA, 1997.
12. J. Burkholder et al., 1993.
13. A. J. Underwood and E. J. Denley, 1984; P. K. Dayton, 1992.
14. R. T. Paine, 1966.
15. D. Raffaelli and S. Hawkins, 1996; H. A. Mooney et al., 1995.
16. R. T. Paine, J. C. Castilla, and J. Cancino, 1985; D. Raffaelli and S. Hawkins, 1996; J. A. Estes, D. O. Duggins, and G. B. Rathbun, 1989.
17. J. Roughgarden, S. Gaines, and H. Possingham, 1988; S. D. Gaines and J. Roughgarden, 1987.
18. P. K. Dayton, 1992; H. A. Mooney et al., 1995.
19. E. G. Leigh et al., 1984.
20. J. Lewin, 1978; D. Raffaelli and S. Hawkins, 1996.
21. H. A. Mooney et al., 1995.

22. M. L. Reaka-Kudla, 1997; J. W. Wells, 1957.
23. M. L. Reaka-Kudla, 1997.
24. National Research Council, 1995; J. E. Maragos, M. P. Crosby, and J. W. McManus, 1996.
25. J. B. C. Jackson, 1997.
26. J. H. Connell, 1978; National Research Council, 1995; J. E. Maragos, M. P. Crosby, and J. W. McManus, 1996.
27. M. A. Huston, 1985.
28. P. F. Sale, 1980; F. G. Stehli and J. W. Wells, 1971; C. Birkeland, 1990.
29. J.E. Maragos, M. P. Crosby, and J. W. McManus, 1996; J. Ogden, 1989.
30. E. Pennisi, 1997.
31. D. M. Alongi, 1990; N. Coleman, A Gason, and G. Poore, 1997; M. L. Reaka-Kudla, 1997.
32. P. W. Glynn, 1988; K. Fauchald, 1991.
33. K. Sherman et al., 1988; G. Mayer, 1982.
34. F. Mowat, 1996.
35. B. Rygg, 1985; D. F. Boesch, 1982.
36. M. Williamson, 1997.
37. K. Sherman et al., 1988; K. H. Mann and J. R. N. Lazier, 1996.
38. A. P. McGinn, 1998.
39. D. Pauly et al., 1998; P. K. Dayton, 1998.
40. A. P. McGinn, 1998; M. H. Prager and A. D. MacCall, 1993; J. Alheit and E. Hagen, 1997.
41. F. Mowat, 1996; C. Safina, 1998; S. A. Earle, 1995.
42. T. Colborn, D. Dumanoski, and J. P. Myers, 1997.
43. D. Anderson, 1997; T. Smayda, 1992; T. Smayda, 1990.
44. D. Anderson, 1997; E. Culotta, 1992; T. Smayda, 1997a.
45. D. Anderson, 1997; H. Mianzan et al., 1997.

Chapter Five

1. K. H. Mann and J. R. N. Lazier, 1996.
2. S. Pain, 1988; J. Gage and P. Tyler 1991.
3. H.L. Sanders, 1968; L. R. Pomeroy 1992; S. W. Chisholm et al., 1988; S. W. Chisholm, 1992.
4. P. A. Jumars, 1976; M. Angel, 1997.
5. J. Hardy, 1991; Y. Zeitsev, 1992.
6. J. Hardy and C. W. Apts, 1989.
7. K. H. Mann and J. R. N. Lazier, 1996; M. Angel, 1993; M. Angel, 1997; J. Raymont, 1963.
8. T. Hayward, 1993; E. L. Venrick, 1990.
9. J. A. McGowan and P. W. Walker, 1993; M. Angel, 1997; A. C. Pierrot-Bults, 1997; S. J. Giovannoni et al., 1990; T. Villareal, M. Altabet, and K. Culver-Rymsza, 1993.
10. M. Angel, 1993, 1997; R. L. Haedrich, 1996; K. H. Mann and J. R. N. Lazier, 1996.
11. M. Angel, 1997; G. C. Ray, 1991; J. A. McGowan and P. W. Walker, 1993; P. Greenwood, 1992; M. Williamson, 1997; A. C. Pierrot-Bults, 1997.
12. M. Angel, 1993; R. L. Haedrich, 1997.
13. K. H. Mann and J. R. N. Lazier, 1996; J. A. McGowan and P. W. Walker, 1993.
14. G. D. Grice and A. D. Hart, 1962; M. Angel, 1993; K. Banse, 1994; K. H. Mann and J. R. N. Lazier, 1996.

15. J. A. McGowan, 1986; K. H. Mann and J. R. N. Lazier, 1996; M. Angel, 1993; J. A. McGowan and P. W. Walker, 1993.
16. J. A. McGowan, 1986; J. A. McGowan and P. W. Walker, 1993.
17. K. H. Mann and J. R. N. Lazier, 1996.
18. K. H. Mann and J. R. N. Lazier, 1996.
19. P. K. Dayton et al., 1994.
20. R. A. Massom, 1988.
21. P. K. Dayton et al., 1994.
22. M. Angel, 1993; AMAP, 1997; Working Group on the Protection of the Arctic Marine Environment, 1996; A. Clarke and J. A. Crame, 1997; J. E. Winston, 1990.
23. J. Gage and P. Tyler, 1991.
24. R. L. Haedrich, 1996; R. L. Haedrich and N. Merrett, 1990; J. Gage and P. Tyler, 1991.
25. J. A. Koslow, 1997.
26. J. Gage and P. Tyler, 1991; J. Gage, 1997; T. Waters, 1995.
27. National Research Council, 1995; J. Gage and P. Tyler, 1991; B. A. Bennett et al., 1994.
28. J. F. Grassle and N. J. Maciolek, 1992; R. M. May, 1994a; G. C. B. Poore and G. D. F. Wilson, 1993; M. L. Reaka-Kudla, 1997; J. C. Briggs, 1994.
29. P. A. Jumars, 1976; J. Gage and P. Tyler, 1991.
30. National Research Council, 1993; J. Gage and P. Tyler, 1991; M. A. Rex, R. J. Etter, and C. T. Stuart, 1997.
31. J. F. Grassle, 1989; J. F. Grassle and N. J. Maciolek, 1992; J. Gage and P. Tyler, 1991; M. A. Rex, R. J. Etter, and C. T. Stuart, 1997.
32. J. Gage and P. Tyler, 1991; J. Gage, 1997.
33. L. G. Abele and K. Walters, 1979; M. A. Rex, 1981.
34. P. K. Dayton and R. R. Hessler, 1972; J. Gage and P. Tyler, 1991.
35. J. F. Grassle, 1989; P. A. Jumars, 1976.
36. J. Gage and P. Tyler, 1991; C. L. Van Dover, 1996; J. Travis, 1993.
37. J. J. Stegeman, P. J. Kloepper-Sams, and J. W. Farington, 1986.
38. J. Hardy, 1991.
39. A. Aarkrog et al., 1987; W. J. Davis, 1994.
40. T. Appenzeller, 1991.
41. C. L. Van Dover, 1996, pp. 3 and 7.

Chapter Six

1. R. V. Salm with J. R. Clark, 1984. The IUCN is now known as the World Conservation Union.
2. National Research Council, 1995.
3. Mellman Group, 1996.
4. W. Eichbaum et al., 1996.
5. Information about Channel Islands National Marine Sanctuary is available on the Internet at <http://www.cinms.nos.noaa.gov>.
6. J. H. Steele et al., 1989.
7. State of Oregon Ocean Policy Advisory Council, 1994; P. Wilson and D. P. Wheeler, 1997.
8. D. Pauly, 1997.
9. J. Caddy 1997.
10. C. E. Curtis, 1990
11. B. Thorne-Miller, 1992.
12. T. Colborn, D. Dumanoski, and J. P. Myers, 1997.
13. U.S. Congress, Office of Technology Assessment, 1986.

14. R. Wilson and E. A. C. Crouch, 1987.
15. J. Cairns Jr., 1986; J. Cairns Jr. and J. R. Pratt, 1989; R. Hilborn and D. Ludwig, 1993.
16. CalCOFI Committee, 1990; the quote is from C. M. Duarte, J. Cebrian, and N. Marba, 1992, p. 190.
17. S. L. Pimm 1997.
18. R. Costanza et al., 1997; S. L. Pimm, 1997; B. Holmes, 1997.
19. T. Beatley, 1991.
20. J. Cairns Jr., 1989.

Chapter Seven

1. C. de Klemm and C. Shine, 1993.
2. The following references apply to the entire discussion of various treaties and agreements in this chapter but are cited only this once: C. de Klemm and C. Shine, 1993; L. Kimball, 1995; R. L. Wallace, 1997.
3. World Commission on Environment and Development, 1987, p. 264.
4. UNEP, 1995.
5. S. M. Wells and J. G. Barzdo, 1991.
6. C. Wilkinson and B. Salvat, 1997.
7. B. Hulshoff and W. P. Gregg, 1985.
8. IMO, 1990.
9. H.L. Windom, 1991.
10. D. R. Downes and B. Van Dyke, 1998.
11. T. E. Wirth, 1995, p. 30.

Chapter Eight

1. F. S. Chapin III et al., 1997, p. 500; N. Myers, 1993, p. 75.
2. Robert May is a noted English scientist. He expressed this opinion at the National Academy of Sciences conference "Nature and Human Society: The Quest for a Sustainable World," October 27–30, 1997, Washington, D.C.
3. E. O. Wilson, 1984.
4. J. Lubchenco, 1998, p. 491.
5. E. O. Wilson, 1984, p. 119.
6. David Suzuki is producer of the Canadian TV series "The Nature of Things" and director of the Suzuki Foundation,which supports environmental projects. He spoke at the conference "Nature and Human Society: The Quest for a Sustainable World," October 27–30, 1997, Washington, D.C.
7. Peter Raven is a botanist and director of the Missouri Botanical Garden; he spoke at "Nature and Human Society: The Quest for a Sustainable World," October 27–30, 1997, Washington, D.C.
8. D. Freestone, 1991; T. Jackson and P. J. Taylor, 1992; R. Wilder, 1997.
9. N. Myers, 1993, p. 77.
10. Norman Myers, an independent English scientist who is well known for his work on biodiversity, spoke at "Nature and Human Society: The Quest for a Sustainable World," October 27–30, 1997, Washington, D.C.
11. Paul Hawken is a business innovator, author, and consultant in Sausalito, California. He spoke at "Nature and Human Society: The Quest for a Sustainable World," October 27–30, 1997, Washington, D.C.
12. P. Bierman-Lytle, 1995; see also the discussion of clean production technology in Chapter 6.
13. N. J. Todd and J. Todd, 1994, p. xvii.

14. Jerry Schubel, director of the New England Aquarium in Boston, spoke at "Nature and Human Society: The Quest for a Sustainable World," October 27–30, 1997, Washington, D.C. See also the discussion of the endangered species approach in Chapter 7.

15. T. E. Wirth, 1995, p. 31.

16. I. Serageldin and R. Barrett, 1995, p. v.

17. Press release, September 28, 1997, Thessaloniki, Greece, announcing "Religion, Science, and the Environment Symposium II: The Black Sea in Crisis," September 20–28, 1997.

Bibliography

Aarkrog, A., S. Boelskifte, H. Dahlgaard, S. Duniec, L. Hallstadius, E. Helm, and J. N. Smith. 1987. Technetium-99 and cesium-134 as long distance tracers in Arctic waters. *Estuarine, Coastal, and Shelf Science* 24:637–647.

Abbott, I., and G. Hollenberg. 1976. *Marine Algae of California.* Stanford University Press, Stanford, Calif.

Abele, L. G., and K. Walters. 1979. The stability-time hypothesis: Reevaluation of the data. *American Naturalist* 114:559–568.

Agardy, T. 1988. *A Status Report on the Workings of the Joint U.S./Canada ad hoc Selection Panel for Acadian Boreal Biosphere Reserve Nomination.* Prepared for the U.S. MAB Directorate on Biosphere Reserves, U.S. Department of State, Washington, D.C.

Alheit, J., and E. Hagen. 1997. Long-term climate forcing of European herring and sardine populations. *Fisheries Oceanography* 6:130–139.

Alongi, D. M. 1990. The ecology of tropical soft-bottom benthic ecosystems. *Oceanography and Marine Biology Annual Review* 28:381–496.

AMAP (Arctic Monitoring and Assessment Programme). 1997. *Arctic Pollution Issues: A State of the Arctic Environment Report.* AMAP, Oslo, Norway.

Ames, B., R. Magaw, and L. S. Gold. 1987. Ranking possible carcinogenic hazards. *Science* 236:271–279.

Anderson, D. 1997. Turning back the harmful red tide. *Nature* 388:513–514.

Anderson, D., and A. White. 1992. Marine biotoxins at the top of the food chain. *Oceanus* 35:55–61.

Angel, M. 1993. Biodiversity of the pelagic ocean. *Conservation Biology* 7:760–772.

Angel, M. 1997. Pelagic biodiversity. Chap. 3 (pp. 35–68) in R. Ormond, J. Gage, and M. Angel (eds.), *Marine Biodiversity: Patterns and Processes.* Cambridge University Press, Cambridge, England.

Angermeier, P. L., and J. R. Karr. 1994. Biological integrity versus biological diversity as policy directives. *BioScience* 44:690–697.

Appenzeller, T. 1991. Fire and ice under the deep-sea floor. *Science* 252:1790–1792.

Ayala, F. J., and J. W. Valentine. 1979. Genetic variability in the pelagic environment: A paradox. *Ecology* 60:24–29.

Bailey, C. 1988. The political economy of fisheries development in the third world. *Agricultural and Human Values* 5:35–48.

Bak, R. P. M. 1987. Effects of chronic oil pollution on a Caribbean coral reef. *Marine Pollution Bulletin* 18:534–539.

Banse, K. 1990. Does iron really limit phytoplankton production in the offshore subarctic Pacific? *Limnology and Oceanography* 35:772–775.

Banse, K. 1994. Grazing and zooplankton production as key controls of phytoplankton production in the open ocean. *Oceanography* 7:13–20.

Barbault, R., and S. Sastrapradja. 1995. Generation, maintenance, and loss of biodiversity. Chap. 4 (pp. 193–274) in V. Heywood (ed.), *UNEP: Global Biodiversity Assessment.* Cambridge University Press, Cambridge, England.

Barnett, M. A. 1983. Species structure and temporal stability of mesopelagic fish assemblages in the central gyres of the North and South Pacific Ocean. *Marine Biology* 74:245–256.

Baskin, Y. 1994. Ecosystem function of biodiversity. *BioScience* 44:657–660.

Beatley, T. 1991. Protecting biodiversity in coastal environments: Introduction and overview. *Coastal Management* 19:1–19.

Beebe, W. 1951. *Half Mile Down.* Duell, Sloan, and Pearce, New York.

Belsky, M. 1986. Legal constraints and options for total ecosystem management of large marine ecosystems. Chap. 11 (pp. 241–262) in K. Sherman and L. Alexander (eds.), *Variability and Management of Large Marine Ecosystems.* AAAS Selected Symposia Series, No. 99. Westview Press, Boulder, Colo.

Ben-Eliahu, M. N., and U. N. Safriel. 1982. A comparison between species diversities of polychaetes from tropical and temperate structurally similar rocky intertidal habitats. *Journal of Biogeography* 9:371–390.

Ben-Eliahu, M. N., U. N. Safriel, and S. Ben-Tuvia. 1988. Environmental stability is low where polychaete species diversity is high: Quantifying tropical vs. temperate within-habitat features. *Oikos* 52:255–273.

Bennett, B. A., C. R. Smith, B. Glaser, and H. L. Maybaum. 1994. Faunal community structure of a chemoautotrophic assemblage on whale bones in the deep northeast Pacific Ocean. *Marine Ecology Progress Series* 3:205–223.

Berra, T. 1997. Some twentieth-century fish discoveries. *Environmental Biology of Fishes* 50:1–12.

Bierman-Lytle, P. 1995. Values in urban infrastructure and development: panelist presentation. Pp. 18–19 in I. Serageldin and R. Barrett (eds.), *Ethics and Spiritual Values: Promoting Environmentally Sustainable Development.* Environmentally Sustainable Development Proceedings Series, No. 12. World Bank, Washington, D.C.

Bigg, G. 1996. *The Oceans and Climate.* Cambridge University Press, Cambridge, England.

Birkeland, C. 1990. Geographic comparisons of coral-reef community processes. In *Proceedings of the Sixth International Coral Reef Symposium,* 1988. Queensland, Australia.

Blaxter, J. H. S., and C. C. Ten Hallers-Tjabbes. 1992. The effect of pollutants on sensory systems and behaviour of aquatic animals. *Netherlands Journal of Aquatic Ecology* 26:43–58.

Boesch, D. F. 1974. Diversity, stability, and response to human disturbance in estuarine ecosystems. Pp. 109–114 in *Structure, Functioning, and Management of Ecosystems: Proceedings of the First International Congress of Ecology, The Hague, The Netherlands, September 8–14, 1974.* Pudoc, Wageningen, Netherlands.

Boesch, D. F. 1974. Ecosystem consequences of alterations of benthic community structure and function in the New York Bight region. Pp. 543–568 in G. Mayer (ed.), *Ecological Stress and the New York Bight: Science and Management.* Estuarine Research Federation, Columbia, S.C.

Bossi, R., and G. Cintron. 1990. *Mangroves of the Wider Caribbean: Toward Sustainable Management.* United Nations Environment Programme, Nairobi, Kenya.

Botsford, L. W., J. C. Castilla, and C. H. Peterson. 1997. The management of fisheries and marine ecosystems. *Science* 277:509–515.

Boyle, E. A. 1990. Quaternary deep water paleooceanography. *Science* 249:863–869.

Brand, L. E. 1981. Genetic variability in reproduction rates in marine phytoplankton populations. *Estuaries* 35:1117–1127.

Briggs, J. C. 1994. Species diversity: Land and sea compared. *Systematic Biology* 43:130–135.

Broecker, W. S. 1990. Comment on "Iron deficiency limits phytoplankton growth in Antarctic waters" by John H. Martin et al. *Global Biogeochemical Cycles* 4:3–4.

Broecker, W. S., and G. H. Denton. 1990. What drives glacial cycles? *Scientific American,* January, 49–56.

Brown, L. R., W. U. Chandler, C. Flavin, C. Pollock, S. Postel, L. Starke, and E. C. Wolf. 1985. *State of the World 1985.* W. W. Norton and Company, New York.

Brusca, R., and G. Brusca. 1990. *Invertebrates.* Sinauer Associates, Sunderland, Mass.

Bryant, P. J. 1998. *Biodiversity and Conservation: A Hypertext Book.* Internet <http://darwin.bio.uci.edu>. University of California, Irvine. (Access through <www.yahoo.com>, search category: biodiversity and conservation.)

Bucklin, A. 1986. The genetic structure of zooplankton populations. Pp. 33–41 in *Pelagic Biogeography: Proceedings of an International Conference, the Netherlands, 29 May–5 June 1985.* UNESCO Technical Papers in Marine Science, No. 49. United Nations Educational, Scientific, and Cultural Organization, Paris.

Burkholder, J., E. Noga, C. Hobbs, and H. Glasgow Jr. 1992. New "phantom" dinoflagellate is the causative agent of major estuarine fish kills. *Nature* 358:407–410.

Burton, R. S. 1983. Protein polymorphisms and genetic differentiation of marine invertebrate populations. *Marine Biology Letters* 4:193–206.

Caddy, J. 1997. Checks and balances in the management of marine fish stocks: Organizational requirements for a limited reference point approach. *Fisheries Research* 30:1–15.

Cairns, J., Jr. 1986. Emergence of integrative environmental management. Pp. 232–241 in C. Kou, and T. Younos (eds.), *Effects of Upland and Shoreline Land Use on the Chesapeake Bay.* Virginia Polytechnic Institute and State University, Blacksburg.

Cairns, J., Jr. 1987. Can the global loss of species be stopped? *Speculations in Science and Technology* 11:189–196.

Cairns, J., Jr. 1989. Restoring damaged ecosystems: Is predisturbance condition a viable option? *Environmental Professional* 11:152–159.

Cairns, J., Jr. 1990. The prediction, validation, monitoring, and mitigation of anthropogenic effects upon natural systems. *Environmental Auditor* 2:19–25.

Cairns, J., Jr., and J. R. Pratt. 1989. The scientific basis of bioassays. *Hydrobiologia* 188/189:5–20.

Cairns, J., Jr., and J. R. Pratt. 1990. Biotic impoverishment: Effects of anthropogenic stress. Chap. 24 (pp. 495–505) in G. Woodwell (ed.), *The Earth in Transition: Patterns and Processes of Biotic Impoverishment.* Cambridge University Press, Cambridge, England.

CalCOFI Committee (ed.). 1990. Ocean outlook: Global change and the marine environment. *California Cooperative Oceanic Fisheries Investigations Reports* 31:25–27.

Carlson, C. 1988. NEPA and the conservation of biological diversity. *Environmental Law* 19:15–36.

Carlton, J. 1993. Neoextinctions of marine invertebrates. *American Zoologist* 33:499–509.

Carlton, J., and J. Geller. 1993. Ecological roulette: The global transport of nonindigenous marine organisms. *Science* 261:78–80.

Chapin, F. S., III, et al. 1997. Biotic control over the functioning of ecosystems. *Science* 277:500–504.

Charlson, R., J. Langner, H. Rodhe, C. B. Leovy, and S. G. Warren. 1991. Perturbation of the Northern Hemisphere radiative balance by backscattering from anthropogenic sulfate aerosols. *Tellus* 43A-B:152–163.

Charlson, R. J., J. E. Lovelock, M. O. Andreae, and S. G. Warren. 1987. Oceanic phytoplankton, atmospheric sulfur, cloud albedo, and climate. *Nature* 326:655–661.

Chisholm, S. W. 1992. What limits phytoplankton growth? *Oceanus*35:36–46.

Chisholm, S. W., R. J. Olson, E. R. Zettler, R. Goericke, J. B. Waterbury, and N. A. Wetschmeyer. 1988. A novel free-living prochlorophyte abundant in the oceanic euphotic zone. *Nature* 334:340–343.

Cho, B., and F. Azam. 1988. Major role of bacteria in biogeochemical fluxes in the ocean's interior. *Nature* 332:441–442.

Clarke, A., and J. A. Crame. 1997. Diversity, latitude and time: Patterns in the shallow sea. Chap. 6 (pp. 122–147) in R. Ormond, J. Gage, and M. Angel (eds.), *Marine Biodiversity: Patterns and Processes.* Cambridge University Press, Cambridge, England.

Cohen, J. E., C. Small, A. Mellinger, J. Gallup, and J. Sachs. 1997. Estimates of coastal populations. *Science* 278:1211–1212.

Colborn, T., D. Dumanoski, and J. P. Myers. 1997. *Our Stolen Future.* Penguin Group, Plume, New York.

Coleman, N., A. Gason, and G. Poore. 1997. High species richness in the shallow marine waters of southeast Australia. *Marine Ecology Progress Series* 154:17–26.

Colwell, R. R. 1983. Biotechnology in the marine sciences. *Science* 222:19–24.

Colwell, R. R. 1997. Microbial biodiversity and biotechnology. Chap. 19 (pp. 279–288) in M. L. Reaka-Kudla, D. E. Wilson, and E. O. Wilson (eds.), *Biodiversity II: Understanding and Protecting Our Biological Resources.* National Academy Press, Joseph Henry Press, Washington, D.C.

Connell, J. H. 1978. Diversity in tropical rain forests and coral reefs. *Science* 199:1302–1310.

Connell, J. H. 1983. On the prevalence and relative importance of interspecific competition: Evidence from field experiments. *American Naturalist* 122:661–696.

Connor, J., and C. Baxter. 1989. *Kelp Forests.* Monterey Bay Aquarium Foundation, Monterey, Calif.

Conversi, A., and J. McGowan. 1994. Natural versus human-caused variability of water clarity in the Southern California Bight. *Limnology and Oceanography* 39:632–648.

Costanza, R., W. M. Kemp, and W. R. Boynton. 1993. Predictability, scale, and biodiversity in coastal and estuarine ecosystems: Implications for management. *Ambio* 22:88–96.

Costanza, R., et al. 1997. The value of the world's ecosystem services and natural capital. *Nature* 387:253–260.

Council on Environmental Quality. 1988. *Environmental Quality, 1986:*

Seventeenth Annual Report of the Council on Environmental Quality. U.S. Government Printing Office, Washington, D.C.

Culotta, E. 1992. Red menace in the world's oceans. *Science* 257:1476–1477.

Culotta, E. 1994. Is marine biodiversity at risk? *Science* 263:918–920.

Curtis, C. 1988. Environmental conditions and trends in marine and near-coastal waters. Testimony before the Subcommittee on Environmental Protection of the Senate Committee on Environment and Public Works, April 20.

Curtis, C. E. 1990. Protecting the oceans. *Oceanus* 3:19–22.

Curtis, C. E. 1993. International ocean protection agreements: What is needed? Chap. 14 (pp. 187–197) in J. M. Van Dyke, D. Zaelke, and G. Hewison (eds.), *Freedom for the Seas in the Twenty-First Century: Ocean Governance and Environmental Harmony.* Island Press, Washington, D.C.

Davies, A. 1990. Taking a cool look at iron. *Nature* 345:114–115.

Davis, W. J. 1993. The need for a new global ocean governance system. Chap. 12 (pp. 147–170) in J. M. Van Dyke, D. Zaelke, and G. Hewison (eds.), *Freedom for the Seas in the Twenty-First Century: Ocean Governance and Environmental Harmony.* Island Press, Washington, D.C.

Davis, W. J. 1994. Contamination of coastal versus open ocean surface waters, a brief meta-analysis. *Marine Pollution Bulletin* 26:128–134.

Dayton, P. K. 1984. Processes structuring some marine communities: Are they general? Chap. 12 (pp. 181–197) in D. R. Strong, Jr. (ed.), *Ecological Communities: Conceptual Issues and the Evidence.* Princeton University Press, Princeton, N.J.

Dayton, P. K. 1992. Community landscape: Scale and stability in hard bottom marine communities. Pp. 289–332 in P. Giller, A. Hildrew, and D. Raffaelli (eds.), *Aquatic Ecology: Scale, Pattern, and Process.* Blackwell Scientific Publications, Oxford, England.

Dayton, P. K. 1998. Reversal of the burden of proof in fisheries management. *Science* 279:821–822.

Dayton, P. K., and R. R. Hessler. 1972. Role of biological disturbance in maintaining diversity in the deep sea. *Deep-Sea Research* 19:199–208.

Dayton, P. K., B. J. Mordida, and F. Bacon. 1994. Polar marine communities. *American Zoologist* 34:90–99.

Dayton, P. K., M. T. Tegner, P. B. Edwards, and K. L. Riser. 1998. Sliding baselines, ghosts, and reduced expectations in kelp forest communities. *Ecological Applications* 8:309–322.

Dayton, P. K., S. Thrush, M. T. Agardy, and R. Hofman. 1995. Environmental effects of marine fishing. *Aquatic Conservation: Marine and Freshwater Ecosystems* 5:205–232.

Deegan, L. 1993. Nutrient and energy transport between estuaries and coastal marine ecosystems by fish migration. *Canadian Journal of Fisheries and Aquatic Science* 50:74–79.

de Klemm, C., with C. Shine. 1993. *Biological Diversity Conservation and the Law: Legal Mechanisms for Conserving Species and Ecosystems.* Environmental Policy and Law Paper No. 29. IUCN—the World Conservation Union, Gland, Switzerland.

Deming, J., and P. Yager. 1992. Natural bacterial assemblages in deep-sea sediments: Towards a global view. Pp. 11–27 in G. T. Rowe and V. Pariente (eds.), *Deep-Sea Food Chains and the Global Carbon Cycle.* Kluwer Academic Publishers, Dordrecht, Netherlands.

Downes, D. R., and B. Van Dyke. 1998. *Fisheries Conservation and Trade Rules: Ensuring that Trade Law Promotes Sustainable Fisheries.* Center for International Environmental Law, and Greenpeace, Washington, D.C.

Duarte, C. M., J. Cebrian, and N. Marba. 1992. Uncertainty of detecting sea change. *Nature* 356:190.
Dunbar, M. J. 1968. *Ecological Development in Polar Regions: A Study in Evolution.* Prentice Hall, Englewood Cliffs, N.J.
Dymond, J. 1992. Particles in the ocean. *Oceanus* 35:60–67.
Earle, S. A. 1991. Sharks, squids, and horseshoe crabs—The significance of marine biodiversity. *BioScience* 41:506–509.
Earle, S. A. 1995. *Sea Change: A Message of the Oceans.* Addison-Wesley, Reading, Mass.
Ehler, C. N. 1994. Implementing an integrated, continuous management process for the Florida Keys National Marine Sanctuary. Paper presented at the Third Annual Symposium of the Ocean Governance Study Group, April 9–13, 1994. Lewes, Del.
Eichbaum, W., M. Crosby, M. Agardy, and S. Laskin. 1996. The role of marine and coastal protected areas in the conservation and sustainable use of biological diversity. *Oceanography* 9:60–70.
Emery, A. R. 1978. The basis of fish community structure: Marine and freshwater comparisons. *Environmental Biology and Fisheries* 3:33–47.
EPA (Environmental Protection Agency). Office of Marine and Estuarine Protection. 1986. Near coastal waters: Strategic options paper. Unpublished report. EPA, Washington, D.C.
EPA (Environmental Protection Agency). 1997. *Incidence and Severity of Sediment Contamination in Surface Waters of the United States.* Vol. 1 of *National Sediment Quality Survey: A Report to Congress on the Extent and Severity of Sediment Contamination in Surface Waters of the United States.* EPA-823-R-97-006. EPA, Washington, D.C.
Estes, J. A., D. O. Duggins, and G. B. Rathbun. 1989. The ecology of extinctions in kelp forest communities. *Conservation Biology* 3:252–264.
Etter, R. J., and J. F. Grassle. 1992. Patterns of species diversity in the deep sea as a function of sediment particle size diversity. *Nature* 360:576–578.
Farnsworth, E. J., and A. M. Ellison. 1997. The global conservation status of mangroves. *Ambio* 26:328–334.
Fauchald, K. 1991. Marine shallow-water soft benthos: Dynamics and conservation issues. Unpublished note.
Fenical, W. 1982. Natural products chemistry in the marine environment. *Science* 215:923–928.
Fenical, W. 1996. Marine biodiversity and the medicine cabinet: The status of new drugs from marine organisms. *Oceanography* 9:23–27.
Fisher, D., J. Ceruso, T. Mathew, and M. Oppenheimer. 1988. *Polluted Coastal Waters: The Role of Acid Rain.* Environmental Defense Fund, New York.
Fortes, M. D. 1988. Mangrove and seagrass beds of East Asia: Habitats under stress. *Ambio* 17:207–213.
Foster, N., and M. H. Lemay (eds.). 1988. *Managing Marine Protected Areas: An Action Plan.* Department of State Publication No. 9673. U.S. Department of State, Man and the Biosphere Program, Washington, D.C.
Fox, M. 1996. *The Boundless Circle: Caring for Creatures and Creation.* Quest Books, Theosophical Publishing House, Wheaton, Ill.
Freestone, D. 1991. The precautionary principle. Chap. 2 (pp. 21–39) in R. Churchill and D. Freestone (eds.), *International Law and Global Climate Change.* Graham & Trotman, London.
Frodeman, R. 1996. The rhetoric of science. *GSA Today,* August, 12–13.
Frost, B. 1987. Grazing control of phytoplankton stock in the open subarctic Pacific Ocean: A model assessing the role of mesozooplankton, particularly the

large calanoid copepods *Neocalanus* spp. *Marine Ecology Progress Series* 39:49–68.

Frost, B. W. 1996. Phytoplankton bloom on iron rations. *Nature* 383:475–476.

Gage, J. 1997. High benthic species diversity in deep-sea sediments: The importance of hydrodynamics. Chap. 7 (pp. 148–177) in R. Ormond, J. Gage, and M. Angel (eds.), *Marine Biodiversity: Patterns and Processes.* Cambridge University Press, Cambridge, England.

Gage, J., and P. Tyler. 1991. *Deep-Sea Biology: A Natural History of Organisms at the Deep-Sea Floor.* Cambridge University Press, Cambridge, England.

Gaines, S. D., and J. Roughgarden. 1987. Fish in offshore kelp forests affect recruitment to intertidal barnacle populations. *Science* 235:479–481.

Gallagher, J. C. 1980. Population genetics of *Skeletonema costatum* (Bacillariophyceae) in Narragansett Bay. *Journal of Phycology* 16:464–474.

Gallagher, R., and B. Carpenter. 1997. Human-dominated ecosystems. *Science* 277:485.

Gargett, A. 1997. The optimal stability "window": A mechanism underlying decadal fluctuations in the North Pacific salmon stocks? *Fisheries Oceanography* 6:109–117.

GESAMP (Joint Group of Experts on the Scientific Aspects of Marine Pollution). 1990. *The State of the Marine Environment.* UNEP Regional Seas Reports and Studies, No. 115. United Nations Environment Programme, Nairobi, Kenya.

Giller, P., A. Hildrew, and D. Raffaelli (eds.). 1992. *Aquatic Ecology: Scale, Pattern, and Process.* Blackwell Scientific Publications, Oxford, England.

Giovannoni, S. J., T. B. Britschgi, C. L. Moyer, and K. C. Field. 1990. Genetic diversity in Sargasso Sea bacterioplankton. *Nature* 345:60–61.

Gitay, H., J. Wilson, and W. Lee. 1996. Species redundancy: A redundant concept. *Journal of Ecology* 84:121–124.

Gittings, S. R., and T. J. Bright. 1988. The *M/V Wellwood* grounding: A sanctuary case study. The science. *Oceanus* 31:36–41.

Glynn, P. W. 1988. El Niño–Southern Oscillation 1982–1983: Nearshore population, community, and ecosystem responses. *Annual Review of Ecology and Systematics* 19:309–345.

Gould, S. J. 1991. On the loss of a limpet. *Natural History* 6:22–27.

Grassle, J. F. 1989. Species diversity in deep-sea communities. *Trends in Ecology and Evolution* 4:12–15.

Grassle, J. F., and Maciolek, N. J. 1992. Deep-sea species richness. *American Naturalist* 139:313–341.

Grassle, J. F., N. J. Maciolek, and J. A. Blake. 1991. Are deep-sea communities resilient? Chap. 17 in G. Woodwell (ed.), *The Earth in Transition: Patterns and Processes of Biotic Impoverishment.* Proceedings of conference held at the Woods Hole Research Center, Woods Hole, Mass., October 1989.

Grassle, J. F., and H. L. Sanders. 1973. Life histories and the role of disturbance. *Deep-Sea Research* 20:643–659.

Grassle, J. P., and J. F. Grassle. 1976. Sibling species in the marine pollution indicator Capitella (Polychaeta). *Science* 192:567–569.

Gray, J. S. 1997. Gradients in marine biodiversity. Chap. 2 (pp. 18–34) in R. Ormond, J. Gage, and M. Angel (eds.), *Marine Biodiversity: Patterns and Processes.* Cambridge University Press, Cambridge, England.

Gray, J. S., and J. M. Bewers. 1996. Towards a scientific definition of the precautionary principle. *Marine Pollution Bulletin* 32:768–771.

Greenwood, P. 1992. Are the major fish faunas well-known? *Netherlands Journal of Zoology* 42:131–138.

Gribbin, J. 1988. The oceanic key to climatic change. *New Scientist,* May 19, 32–33.
Grice, G. D., and A. D. Hart. 1962. The abundance, seasonal occurrence, and distribution of the epizooplankton between New York and Bermuda. *Ecological Monographs* 32:287–309.
Gyllensten, U., and N. Ryman. 1985. Pollution biomonitoring programs and the genetic structure of indicator species. *Ambio* 14:29–31.
Haedrich, R. L. 1996. Deep-water fishes: Evolution and adaptation in the earth's largest living spaces. *Journal of Fish Biology* 49:40–53.
Haedrich, R. L., and N. Merrett. 1990. Little evidence for faunal zonation or communities in deep sea demersal fish faunas. *Progress in Oceanography* 24:239–250.
Hallers-Tjabbes, C., J. Kemp, and J. Boon. 1994. Imposex in whelks (*Buccinum undatum*) from the open North Sea: Relation to shipping traffic intensities. *Marine Pollution Bulletin* 28:311–313.
Hammel, M., A. Jansson, and B. Jansson. 1993. Diversity change and sustainability: Implications for fisheries. *Ambio* 22:97–102.
Harbison, G. 1992. The gelatinous inhabitants of the ocean interior. *Oceanus* 35:18–23.
Hardy, J. 1991. Where the sea meets the sky. *Natural History* 5:59–65.
Hardy, J., and C. W. Apts. 1989. Photosynthetic carbon reduction: High rates in the sea-surface microlayer. *Marine Biology* 101:411–417.
Hart, S. 1989. Food chains: The carbon link. *Science* 136:168–170.
Hawksworth, D. L., and M. T. Kalin-Arroyo. 1995. Magnitude and distribution of biodiversity. Chap. 3 (pp. 107–191) in V. Heywood, *UNEP: Global Biodiversity Assessment.* Cambridge University Press, Cambridge, England.
Hay, M. E., and W. Fenical. 1996. Chemical ecology and marine biodiversity: Insights and products from the sea. *Oceanography* 9:10–20.
Hayden, B. P., G. C. Ray, and R. Dolan. 1984. Classification of coastal and marine environments. *Environmental Conservation* 11:199–207.
Hayward, T. 1993. The rise and fall of *Rhizosolenia. Nature* 363:675–676.
Heap, J. A., and M. W. Holdgate. 1986. The Antarctic Treaty System as an environmental mechanism—An approach to environmental issues. Pp. 195–210 in National Research Council, *Antarctic Treaty System: An Assessment.* National Academy Press, Washington, D.C.
Hedgpeth, J. 1993. Foreign invaders. *Science* 261:34–35.
Hessler, R. R., and H. L. Sanders. 1967. Faunal diversity in the deep sea. *Deep-Sea Research* 14:65–78.
Hilborn, R., and D. Ludwig. 1993. The limits of applied ecological research. *Ecological Applications* 3:550–552.
Holdgate, M. 1990. Biological diversity: Why do we need it? *IUCN Bulletin* 21:27.
Holmes, B. 1997. Don't ignore nature's bottom line. *New Scientist* 30:1–15.
Hulshoff, B., and W. P. Gregg. 1985. Biosphere reserves: Demonstrating the value of conservation in sustaining society. *Parks* 10:2–5.
Huston, M. A. 1985. Patterns of species diversity on coral reefs. *Annual Review of Ecology and Systematics* 16:149–177.
IMO (International Maritime Organization). 1990. *London Dumping Convention: The First Decade and Beyond (Provisions of the Convention on the Prevention of Marine Pollution by Dumping of Wastes and Other Matter, 1972, and Decisions Made by the Consultative Meeting of Contracting Parties, 1975–1989).* LDC 13/Inf.9. IMO Secretariat, London.
IUCN (International Union for the Conservation of Nature). 1987. *Elements of the*

IUCN Coastal and Marine Conservation Programme. IUCN, Gland, Switzerland.

Jackson, J. B. C. 1994. Constancy and change of life in the sea. *Philosophical Transactions of the Royal Society of London,* ser. B, 334:55–60.

Jackson, J. B. C. 1997. Reefs since Columbus. *Coral Reefs* 16: S23–S32 (suppl.).

Jackson, J. B. C., and K. W. Kaufmann. 1987. *Diadema antillarum* was not a keystone predator in cryptic reef environments. *Science* 235:687–689.

Jackson, T., and P. J. Taylor. 1992. The precautionary principle and the prevention of marine pollution. *Chemistry and Ecology* 7:123–134.

Jameson, S. C., J. W. McManus, and M. D. Spalding. 1995. *State of the Reefs: Regional and Global Perspectives.* Background paper of the International Coral Reef Initiative executive secretariat. National Oceanic and Atmospheric Administration, Silver Spring, Md.

Johannes, R. E. 1978. Traditional marine conservation methods in Oceania and their demise. *Annual Review of Ecology and Systematics* 9:349–364.

Johnson, K. H., K. A. Vogt, H. J. Clark, O. J. Schmitz, and D. J. Vogt. 1996. Biodiversity and the productivity and stability of ecosystems. *TREE* 11:373–377.

Johnston, P., and R. Stringer. 1994. . . . *And Then There Were None: Cod Fisheries and Environmental Change.* Greenpeace, Amsterdam, Netherlands.

Jones, J., and J. Reynolds. 1997. Effects of pollution on reproductive behaviour of fishes. *Reviews in Fish Biology and Fisheries* 7:463–464.

Jumars, P. A. 1976. Deep-sea species diversity: Does it have a characteristic scale? *Journal of Marine Research* 34:217–246.

Kaiser, J. 1995. Of whales and ocean warming: A plan to sound out the sea's temperature may be back on course. *Science News.* 147:350–351.

Karentz, D. 1992. Ozone depletion and UV-B radiation in the Antarctic—Limitations to ecological assessment. *Marine Pollution Bulletin* 25:231–232.

Keller, M. 1989. Dimethyl sulfide production and marine phytoplankton: The importance of species composition and cell size. *Biological Oceanography* 6:375–382.

Kerr, R. 1983. Are the ocean's deserts blooming? *Science* 220:397–398.

Kerr, R. A. 1991. An about-face found in the ancient ocean. *Science* 253:1359–1360.

Kimball, L. 1995. The United Nations Convention on the Law of the Sea: A framework for marine conservation. Part 1 (pp. 1–120) in *The Law of the Sea: Priorities and Responsibilities in Implementing the Convention.* IUCN—the World Conservation Union, Gland, Switzerland.

Kime, D. E. 1995. The effects of pollution on reproduction in fish. *Reviews in Fish Biology and Fisheries* 5:52–96.

Kingston, P. F. 1987. Field effects of platform discharges on benthic macrofauna. *Philosophical Transactions of the Royal Society of London,* ser. B, 316:545–565.

Kinzig, A. P., and R. H. Socolow. 1994. Human impacts on the nitrogen cycle. *Physics Today,* November, 24–31.

Kirchner, J. W. 1989. The Gaia hypothesis: Can it be tested? *Reviews of Geophysics* 27:223–235.

Klerks, P., and J. S. Levinton. 1989. Rapid evolution of resistance to extreme metal pollution in a benthic oligochaete. *Biological Bulletin* 176:135–141.

Koslow, J. A. 1997. Seamounts and the ecology of deep-sea fisheries. *American Scientist* 85:168–176.

Langston, W. J., N. D. Pope, and G. R. Burt. 1992. *Impact of Discharges on Metal*

Levels in Biota of the West Cambria Coast. Plymouth Marine Laboratory Report 1992. Plymouth Marine Laboratory, Plymouth, England.

Leigh, E. G., Jr., R. T. Paine, J. F. Quinn, and T. H. Suchanek. 1984. Wave energy and intertidal productivity. *Proceedings of the National Academy of Sciences, U.S.A.* 84:1314–1318.

Lewin, J. 1978. The world of the razor-clam beach. *Pacific Search,* April, 12–13.

Lewontin, R. C. 1990. Fallen angels. *New York Review of Books* 37:3–7.

Lindley, D. 1988. Is the Earth alive or dead? *Nature* 332:483–484.

Lippson, A., and R. Lippson. 1984. *Life in the Chesapeake Bay.* Johns Hopkins University Press, Baltimore, Md.

Livingston, H. D., S. L. Kupferman, V. T. Bowen, and R. M. Moore. 1984. Vertical profile of artificial radionuclide concentrations in the central Arctic Ocean. *Geochimica and Cosmochimica Acta* 48:2195–2203.

Livingstone, D., P. Donkin, and C. Walker. 1992. Pollutants in marine ecosystems: An overview. Chap. 11 (pp. 235–263) in C. Walker and D. Livingstone (eds.), *Persistent Organic Pollutants in Ecosystems.* Pergamon Press, New York.

Lochte, K. 1992. Bacterial standing stock and consumption of organic carbon in the benthic boundary layer of the abyssal North Atlantic. Pp. 1–10 in G. T. Rowe and V. Pariente (eds.), *Deep-Sea Food Chains and the Global Carbon Cycle.* Kluwer Academic Publishers, Dordrecht, Netherlands.

Lovelock, J. E. 1979. *Gaia: A New Look at Life on Earth.* Oxford University Press, New York.

Lovelock, J. E. 1988. The earth as a living organism. Chap. 56 (pp. 487–489) in E. O. Wilson (ed.), *Biodiversity.* National Academy Press, Washington, D.C.

Lovelock, J. E. 1989. Geophysiology, the science of Gaia. *Reviews in Geophysics* 27:215–222.

Lubchenco, J. 1998. Entering the century of the environment: A new social contract for science. *Science* 279:491–497.

Lugo, A. E., and S. C. Snedaker. 1974. The ecology of mangroves. *Annual Review of Ecology and Systematics* 5:39–64.

Lynch, M. P., and G. C. Ray. 1985. *Diversity of Marine/Coastal Ecosystems.* Paper prepared for the Office of Technology Assessment. U.S. Congress, Office of Technology Assessment, Washington, D.C.

McGinn, A. P. 1998. Promoting sustainable fisheries. Pp. 59–78 in *State of the World 1998.* W. W. Norton and Company, New York.

McGowan, J. A. 1986. The biogeography of pelagic ecosystems. Pp. 191–200 in *Pelagic Biogeography: Proceedings of an International Conference, the Netherlands, 29 May–5 June 1985.* UNESCO Technical Papers in Marine Science, No. 49. United Nations Educational, Scientific, and Cultural Organization, Paris.

McGowan, J. A. 1995. Temporal change in marine ecosystems. Pp. 555–571 in National Research Council, *Global Climate Variability on Decade-to-Century Time Scales.* National Academy Press, Washington, D.C.

McGowan, J. A., D. B. Chelton, and A. Conversi. 1996. Plankton patterns, climate, and change in the California Current. *CalCOFI Reports* 37:45–68.

McGowan, J. A., and P. W. Walker. 1979. Structure in the copepod community of the North Pacific central gyre. *Ecological Monographs* 49:195–226.

McGowan, J. A., and P. W. Walker. 1985. Dominance and diversity maintenance in an oceanic ecosystem. *Ecological Monographs* 55:103–118.

McGowan, J. A., and P. W. Walker. 1993. Pelagic diversity patterns. Chap. 19 (pp. 203–214) in R. Ricklefs and D. Schluter (eds.), *Species Diversity in Ecological Communities.* University of Chicago Press, Chicago.

McKibben, W. 1989. The end of nature. *New Yorker,* September 11, 47–105.

McLaughlin, J. F., and J. Roughgarden. 1993. Species interactions in space. Chap. 8 (pp. 89–98) in R. Ricklefs and D. Schluter (eds.), *Species Diversity in Ecological Communities*. University of Chicago Press, Chicago.

McLusky, D. S. 1981. *The Estuarine Ecosystem*. John Wiley and Sons, New York.

McNeely, J. A. 1988. *Economics and Biological Diversity*. International Union for the Conservation of Nature and Natural Resources, Gland, Switzerland.

McNeely, J. A., K. R. Miller, W. V. Reid, R. A. Mittermeier, and T. B. Werner. 1990. *Conserving the World's Biological Diversity*. International Union for the Conservation of Nature and Natural Resources, World Resources Institute, Conservation International, World Wildlife Fund–U.S., and World Bank, Washington, D.C.

Malakoff, D. 1997. Extinction on the high seas. *Science* 277:486–488.

Malakoff, D. 1998. Atlantic salmon spawn fight over species protection. *Science* 279:800.

Manheim, B. S., Jr. 1988. *On Thin Ice: The Failure of the National Science Foundation to Protect Antarctica*. Environmental Defense Fund, New York.

Mann, K. H., and J. R. N. Lazier. 1996. *Dynamics of Marine Ecosystems: Biological-Physical Interactions in the Oceans*. 2nd ed. Blackwell Scientific Publications, Cambridge, England.

Manville, A. M. 1988. Tracking plastic in the Pacific. *Defenders*, November–December, 10–15.

Maragos, J. E., M. P. Crosby, and J. W. McManus. 1996. Coral reefs and biodiversity: A critical and threatened relationship. *Oceanography* 9:83–99.

Margalef, R. 1968. *Perspectives in Ecological Theory*. University of Chicago Press, Chicago.

Margalef, R., and M. Estrada. 1981. On upwelling, eutrophic lakes, the primitive biosphere, and biological membranes. Pp. 522–529 in *Coastal Upwelling*, vol. 1 of F. A. Richards (ed.), *Coastal and Estuarine Science*. American Geophysical Union, Washington, D.C.

Martin, J., W. Broenkow, S. Fitzwater, and R. Gordon. 1990. Yes, it does: A reply to the comment by Banse. *Limnology and Oceanography* 35:775–777.

Martin, J., M. Gordon, and S. Fitzwater. 1990. Iron in Antarctic waters. *Nature* 345:156–158.

Massom, R. A. 1988. The biological significance of open water within the sea ice covers of the polar regions. *Endeavour* (n.s.) 12:21–27.

May, R. M. 1988. How many species are there on Earth? *Science* 241:1441–1448.

May, R. M. 1992. Bottoms up for the oceans. *Nature* 357:278–279.

May, R. M. 1994a. Biological diversity: Differences between land and sea. *Philosophical Transactions of the Royal Society of London*, ser. B, 343:105–111.

May, R. M. 1994b. Conceptual aspects of the quantification of the extent of biological diversity. *Philosophical Transactions of the Royal Society of London*, ser. B, 345:13–20.

Mayer, G. (ed.). 1982. *Ecological Stress and the New York Bight: Science and Management*. Estuarine Research Federation, Columbia, S.C.

Mellman Group. 1996. *Presentation of Findings from a Nationwide Survey and Focus Groups*. SeaWeb, 1731 Connecticut Avenue N.W., 4th Floor, Washington, D.C. 20009.

Menge, B. A. 1992. Community regulations: Under what conditions are bottom-up factors important on rocky shores? *Ecology* 73:755–765.

Merrett, N., and R. Haedrich. 1997. *Deep-Sea Demersal Fish and Fisheries*. Chapman & Hall, London.

Mianzan, H., M. Pajaro, L. Machinandiarena, and F. Cremonte. 1997. Salps: Possible vectors of toxic dinoflagellates. *Fisheries Research* 29:193–197.
Moffat, A. S. 1996. Biodiversity is a boon to ecosystems, not species. *Science* 271: 1497.
Monastersky, R. 1988. Plentiful plankton noticed at last. *Science News* 134:68.
Mooney, H. A., J. Lubchenco, R. Dirzo, and O. E. Sala. 1995. Biodiversity and ecosystem functioning: Ecosystem analysis. Chap. 6 (pp. 327–452) in V. H. Heywood and R. T. Watson (eds.), *UNEP Global Biodiversity Assessment.* Cambridge University Press, Cambridge, England.
Morin, P., and S. Lawler. 1995. Food web architecture and population dynamics: Theory and empirical evidence. *Annual Review of Ecological Systems* 26:505–529.
Morse, D. E., and A. N. C. Morse. 1988. Chemical signals and molecular mechanisms: Learning from larvae. *Oceanus* 31:37–43.
Mowat, F. 1996. *Sea of Slaughter.* Chapters Publishing, Shelburne, Vt.
Mulvaney, K. 1996. Directed kills of small cetaceans worldwide. Chap. 4 (pp. 90–108) in M. P. Simmonds and J. D. Hutchinson (eds.), *The Conservation of Whales and Dolphins.* John Wiley and Sons, New York.
Myers, N. 1979. *The Sinking Ark.* Pergamon Press, Oxford, England.
Myers, N. 1990. Mass extinctions: What can the past tell us about the present and the future? *Palaeogeography, Palaeoclimatology, Palaeoecology* 82: 175–185.
Myers, N. 1992. Population/environment linkages: Discontinuities ahead. *Ambio* 21:116–118.
Myers, N. 1993. Biodiversity and the precautionary principle. *Ambio* 22:74–79.
National Research Council. 1995. *Understanding Marine Biodiversity Science.* National Academy Press, Washington, D.C.
National Research Council. 1997. *Contaminated Sediments in Ports and Waterways: Cleanup Strategies and Technologies.* National Academy Press, Washington, D.C.
National Science Board. 1988. *The Role of the National Science Foundation in Polar Regions.* NSB-87-128. National Science Foundation, Washington, D.C.
National Science Board. 1989. *Loss of Biological Diversity: A Global Crisis Requiring International Solutions.* NSB-89-171. National Science Foundation, Washington, D.C.
Nixon, S. 1997. Prehistoric nutrient inputs and productivity in Narragansett Bay. *Estuaries* 20:253–261.
Nriagu, J. O. 1989. A global assessment of natural sources of atmospheric trace metals. *Nature* 338:47.
Ogden, J. 1989. Marine biological diversity: A strategy for action. *Reef Encounter* 6:5.
Olsen, S., L. Z. Hale, R. DuBois, D. Robadue, and G. Foer. 1989. *Integrated Resources Management for Coastal Environments in the Asia Near East Region.*, University of Rhode Island, International Coastal Resources Management Project, Kingston, R.I.
Ormond, R., J. Gage, and M. Angel (eds.). 1997. *Marine Biodiversity: Patterns and Processes.* Cambridge University Press, Cambridge, England.
Pacific Congress on Marine Science and Technology. 1995. *Proceedings of the PACON Conference on Sustainable Aquaculture 95, June 11–14, 1995, Honolulu, Hawaii.* PACON International, Hawaii Chapter, Honolulu.
Pacific Science Association. Scientific Committee on Coral Reefs. 1988. *Coral Reef Newsletter,* No. 19. Pacific Science Association Secretariat, Bishop Museum, 1525 Bernice Street, Honolulu, Hawaii.

Pain, S. 1988. No escape from the global greenhouse. *New Scientist,* November 12, 38–43.
Pain, S. 1990. On the edge of disaster. *New Scientist,* April 28, 36–37.
Paine, R. T. 1966. Food web complexity and species diversity. American *Naturalist* 100:65–75.
Paine, R. T. 1984. Some approaches to modeling multispecies systems. Pp. 191–207 in R. M. May (ed.), *Exploitation of Marine Communities.* Springer-Verlag, New York.
Paine, R. T., J. C. Castilla, and J. Cancino. 1985. Perturbation and recovery patterns of starfish-dominated intertidal assemblages in Chile, New Zealand, and Washington State. *American Naturalist* 125:679–691.
Pauly, D. 1997. Putting fisheries management back in places. *Reviews in Fish Biology and Fisheries* 7:125–127.
Pauly, D., V. Christensen, J. Dalsgaard, R. Froese, and F. Torres Jr. 1998. Fishing down marine food webs. *Science* 279:860–863.
Pennisi, E. 1997. Brighter prospects for the world's coral reefs? *Science* 277:491–493.
Pianka, E. R. 1966. Latitudinal gradients in species diversity: A review of concepts. *American Naturalist* 100:33–46.
Pianka, E. R. 1988. *Evolutionary Ecology.* Harper & Row, New York.
Pielou, E. C. 1979. *Biogeography.* Wiley-Interscience, New York.
Pierrot-Bults, A. C. 1997. Biological diversity in oceanic macrozooplankton: More than counting species. Chap. 4 in R. Ormond, J. Gage, and M. Angel (eds.), *Marine Biodiversity: Patterns and Processes.* Cambridge University Press, Cambridge, England.
Pimm, S. L. 1984. The complexity and stability of ecosystems. *Nature* 307: 321–326.
Pimm, S. L. 1997. The value of everything. *Nature* 387:231–232.
Pimm, S. L., G. J. Russell, and T. M. Brooks. 1995. The future of biodiversity. *Science* 269:347–350.
Pinet, P. R. 1998. *Invitation to Oceanography.* Jones and Bartlett, Sudbury, Mass.
Platt, T., and S. Sathyendranath. 1992. Scale, pattern, and process in marine ecosystems. Chap. 20 in P. Giller, A. Hildrew, and D. Raffaelli (eds.), *Aquatic Ecology: Scale, Pattern, and Process.* Blackwell Scientific Publications, Oxford, England.
Pomeroy, L. R. 1992. The microbial food web. *Oceanus* 35:28–35.
Pool, R. 1995. Coral chemistry leads to human bone repair. *Science* 267:1772.
Poore, G. C. B., and G. D. F. Wilson. 1993. Marine species richness. *Nature* 362:597–598.
Post, W. M., T. H. Peng, W. R. Emanuel, A. W. King, V. H. Dale, and D. L. DeAngelis. 1990. The global carbon cycle. *American Scientist* 78:310–326.
Prager, M. H., and A. D. MacCall. 1993. Detection of contaminant and climate effects on spawning success of three pelagic fish stocks off southern California: Northern anchovy *Egraulis mordax,* Pacific sardine *Sardinops sagax,* and chub mackerel *Scomber japonicus. Fishery Bulletin,U.S.* 91:310–327.
Raffaelli, D., and S. Hawkins. 1996. *Intertidal Ecology.* Chapman & Hall, London.
Raimondi, P. T., and D. C. Reed. 1993. Determining the spatial extent of ecological impacts caused by local anthropogenic disturbances in coastal marine habitats. Chap. 10 (pp. 179–193) in R. Ricklefs and D. Schluter (eds.), *Species Diversity in Ecological Communities.* University of Chicago Press, Chicago.
Raven, P. H. 1995. The importance of biodiversity. Keynote address for "Symposium on Biodiversity Along the Central California Coast." March 3–5 1995, San Francisco, Calif.
Ray, G. C. 1988. Ecological diversity in coastal zones and oceans. Chap. 4 (pp.

36–50) in E. O. Wilson (ed.), *Biodiversity.* National Academy Press, Washington, D.C.

Ray, G. C. 1991. Coastal-zones biodiversity patterns: Principles of landscape ecology may help explain the processes underlying coastal diversity. *BioScience* 41:490–498.

Ray, G. C. 1997. Do the metapopulation dynamics of estuarine fishes influence the stability of shelf ecosystems? *Bulletin of Marine Science* 60:1040–1049.

Ray, G. C., and J. F. Grassle. 1991. Marine biological diversity: A scientific program to help conserve marine biological diversity is urgently required. *BioScience* 41:453–457.

Raymont, J. 1963. *Plankton and Productivity in the Oceans.* Pergamon Press, Oxford, England.

Reaka-Kudla, M. L. 1997. The global biodiversity of coral reefs: A comparison with rainforests. Chap. 7 (pp. 83–108) in M. L. Reaka-Kudla, D. E. Wilson, and E. O. Wilson (eds.), *Biodiversity II: Understanding and Protecting Our Biological Resources.* National Academy Press, Joseph Henry Press. Washington, D.C.

Reid, W. V., and K. R. Miller. 1989. *Keeping Options Alive: The Scientific Basis for Conserving Biodiversity.* World Resources Institute, Washington, D.C.

Rex, M. A. 1981. Community structure in deep-sea benthos. *Annual Review of Ecology and Systematics* 12:331–353.

Rex, M. A., R. J. Etter, and C. T. Stuart. 1997. Chap. 5 (pp. 94–121) in R. Ormond, J. Gage, and M. Angel (eds.), *Marine Biodiversity: Patterns and Processes.* Cambridge University Press, Cambridge, England.

Rex, M. A., C. T. Stuart, R. R. Hessler, J. A. Allen, H. L. Sanders, and G. D. F. Wilson. 1993. Global-scale latitudinal patterns of species diversity in the deep-sea benthos. *Nature* 365:636–639.

Rice, A., and P. Lambshead. 1992. Patch dynamics in the deep-sea benthos: The role of a heterogeneous supply of organic matter. Chap. 15 in P. Giller, A. Hildrew, and D. Raffaelli (eds.), *Aquatic Ecology: Scale, Pattern, and Process.* Blackwell Scientific Publications, Oxford, England.

Ricketts, E., and J. Calvin. 1962. *Between Pacific Tides,* 3rd ed., rev. Stanford University Press, Stanford, Calif.

Ricklefs, R., and R. Latham. 1993. Global patterns of diversity in mangrove floras. Chap. 20 (pp. 215–229) in R. E. Ricklefs and D. Schluter (eds.), *Species Diversity in Ecological Communities.*University of Chicago Press, Chicago.

Risser, P. 1995. Biodiversity and ecosystem function. *Conservation Biology* 9:742–746.

Ritterhoff, J., and G. Zauke. 1997. Bioaccumulation of trace metals in Greenland Sea copepod and amphipod collectives on board ship: Verification of toxicokinetic model parameters. *Aquatic Toxicology* 40:63–78.

Roemmich, D., and J. McGowan. 1995. Climatic warming and the decline of zooplankton in the California Current. *Science* 267:1324–1326.

Rosenzweig, M., and Z. Abramsky. 1993. How are diversity and productivity related? Chap. 5 (pp. 52–65) in R. Ricklefs, and D. Schluter (eds.), *Species Diversity in Ecological Communities.*University of Chicago Press, Chicago.

Roughgarden, J., S. Gaines, and H. Possingham. 1988. Recruitment dynamics in complex life cycles. *Science* 241:1460–1466.

Roughgarden, J., T. Pennington, and S. Alexander. 1994. Dynamics of the rocky intertidal zone with remarks on generalization in ecology. *Philosophical Transactions of the Royal Society of London,* ser B, 343:79–85.

Rowe, G. T. 1981. Pp. 464–471 in F. A. Richards (ed.), *Coastal and Estuarine Science.* American Geophysical Union, Washington, D.C.

Rozengurt, M., and I. Haydock. 1993. Freshwater flow diversion and its implications for coastal zone ecosystems Pp. 287–295 in *Transactions of the Fifty-Eighth North American Wildlife and Natural Resources Conferences.* Wildlife Management Institute.

Rudyakov, Y. A. 1987. Ecosystems of coastal waters as a component of the biological structure of the ocean. *Oceanology* 27:479–481.

Ruggieri, G. D. 1976. Drugs from the sea. *Science* 194:491–497.

Rygg, B. 1985. Distribution of species along pollution-induced diversity gradients in benthic communities in Norwegian fjords. *Marine Pollution Bulletin* 16:469–474.

Ryther, J., W. Dunstan, K. Tenore, and J. Huguenin. 1972. Controlled eutrophication—Increasing food production from the sea by recycling human wastes. *BioScience* 22:144–152.

Saenger, P., E. J. Hegerl, and J. D. S. Davie (eds.). 1983. Global status of mangrove ecosystems. *Environmentalist* 3, suppl. 3:1–88.

Safina, C. 1995. The world's imperiled fish. *Scientific American* 273:46–53.

Safina, C. 1998. *Song for the Blue Ocean: Encounters Along the World's Coasts and Beneath the Seas.* Henry Holt and Company, New York.

Sale, P. F. 1980. The ecology of fishes on coral reefs. *Oceanography and Marine Biology Annual Review* 18:367–421.

Salm, R. V., with J. R. Clark. 1984. *Marine and Coastal Protected Areas: A Guide for Planners and Managers.* International Union for the Conservation of Nature and Natural Resources, Gland, Switzerland.

Sanders, H. L. 1968. Marine benthic diversity: A comparative study. *American Naturalist* 102:243–282.

Santelices, B. 1990. Patterns of reproduction, dispersal, and recruitment in seaweeds. *Oceanography and Marine Biology Annual Review* 28:177–276.

Sarmiento, J. L. 1991. Slowing the buildup of fossil CO_2 in the atmosphere by iron fertilization: A comment. *Global Biogeochemical Cycles* 5:1–2.

Sarmiento, J., J. Toggweiler, and R. Najjar. 1988. Ocean carbon-cycle dynamics and atmospheric $P\ CO_2$. *Philosophical Transactions of the Royal Society of London,* ser. A, 325:3–21.

Schlesinger, M., and X. Jiang. 1988. The transport of CO_2-induced warming into the ocean: An analysis of simulations by the OSU coupled atmosphere-ocean general circulation model. *Climate Dynamics* 3:1–17.

Schmidt, K. 1997. A drop in the ocean. *New Scientist* 5:40–44.

Schopf, T. J. M., J. B. Fisher, and C. A. F. Smith III. 1978. Is the marine latitudinal diversity gradient merely another example of the species area curve? Pp. 365–386 in B. Battaglia and J. A. Beardmore (eds.), *Marine Organisms: Genetics, Ecology, and Evolution.* Plenum Press, New York.

Serageldin, I., and R. Barrett (eds.). 1995. *Ethics and Spiritual Values: Promoting Environmentally Sustainable Development.* Environmentally Sustainable Development Proceedings Series, No. 12. World Bank, Washington, D.C.

Shane, B. S. 1994. Introduction to ecotoxicology. Chap. 1 (pp. 3–10) in L. G. Cockerham and B. S. Shane (eds.), *Basic Environmental Toxicology.* CRC Press, Boca Raton, Fla.

Sherman, K. 1986. Introduction to parts one and two: Large marine ecosystems as tractable entities for measurement and management. Chap. 1 (pp. 3–8) in K. Sherman and L. Alexander (eds.), *Variability and Management of Large Marine Ecosystems.* AAAS Selected Symposia Series, No. 99. Westview Press, Boulder, Colo.

Sherman, K., J. R. Green, J. R. Goulet, and L. Ejsymont. 1983. Coherence in zooplankton of a large northwest Atlantic ecosystem. *Fishery Bulletin* 81:855–862.

Sherman, K., K. Grosslein, D. Mountain, D. Busch, J. O'Reilly, and R. Theroux. 1988. The continental shelf ecosystem off the northeast coast of the United States. Pp. 279–337 in H. Postma and J. J. Zijlstra (eds.), *Ecosystems of the World,* Vol. 27, *Continental Shelves.* Elsevier, Amsterdam, Netherlands.

Sherr, E. B. 1989. And now, small is plentiful. *Nature* 340:429.

Short, F., and S. Wyllie-Echeverria. 1996. Natural and human-induced disturbance of seagrasses. *Environmental Conservation* 23:17–27.

Simmonds, M., P. Johnston, and M. French. 1993. Organochlorine and mercury contamination in United Kingdom seals. *Veterinary Record* 132:291–295.

Sissenwine, M. P. 1986. Perturbation of a predator-controlled continental shelf ecosystem. Chap. 5 (pp. 55–85) in K. Sherman and L. Alexander (eds.), *Variability and Management of Large Marine Ecosystems.* AAAS Selected Symposia Series, No. 99. Westview Press, Boulder, Colo.

Slovic, P. 1987. Perception of risk. *Science* 236:280.

Smith, C. 1992. Whale falls: Chemosynthesis on the deep seafloor. *Oceanus* 35:74–78.

Smayda, T. 1989. Primary production and the global epidemic of phytoplankton blooms in the sea: A linkage. Pp. 449–483 in E. Cosper, V. Bricelj, and E. Carpenter (eds.), *Novel Phytoplankton Blooms.* Springer-Verlag, New York.

Smayda, T. 1990. Novel and nuisance phytoplankton blooms in the sea: Evidence for a global epidemic. Pp. 29–40 in E. Granéli et al. (eds.), *Toxic Marine Phytoplankton.* Elsevier Science Publishing, New York.

Smayda, T. 1992. Global epidemic of noxious phytoplankton blooms and food chain consequences in large ecosystems. Chap. 13 (pp. 275–307) in K. Sherman, L. Alexander, and B. Gold (eds.), *Food Chains, Yields, Models, and Management of Large Marine Ecosystems.* Westview Press, Boulder, Colo.

Smayda, T. 1997a. Harmful algal blooms: Their ecophysiology and general relevance to phytoplankton blooms in the sea. *Limnology and Oceanography* 42:1137–1153.

Smayda, T. 1997b. What is a bloom? A commentary. *Limnology and Oceanography* 42:1132–1136.

Smith, C. 1992. Whale falls: Chemosynthesis on the deep seafloor. *Oceanus* 35: 74–78.

Smith, P. J., and Y. Fugio. 1982. Genetic variation in marine teleosts: High variability in habitat specialists and low variability in habitat generalists. *Marine Biology* 69:7–20.

Smith, V. K. 1988. Resource evaluation at the crossroads. *Resources* 90:2–6.

Snelgrove, P. 1996. Why care about marine biodiversity? *Sea Technology* 9:93.

Snelgrove, P., J. Grassle, and R. Petrecca. 1992. The role of food patches in maintaining high deep-sea diversity: Field experiments with hydrodynamically unbiased colonization trays. *Limnology and Oceanography* 37:1543–1550.

Sousa, W. P. 1984. The role of disturbance in natural communities. *Annual Review of Ecology and Systematics* 15:353–391.

Sperling, K.-R. 1988. The dangers of risk assessment within the framework of the marine dumping conventions. *Marine Pollution Bulletin* 19:9–10.

Spight, T. T. 1977. Diversity of shallow-water gastropod communities on temperate and tropical beaches. *American Naturalist* 982:1077–1097.

State of Oregon Ocean Policy Advisory Council. 1994. *State of Oregon Territorial Sea Plan.* Oregon Ocean Policy Advisory Council, Portland.

Steele, J. H. 1985. A comparison of terrestrial and marine ecological systems. *Nature* 313:355–358.

Steele, J. H. 1991. Marine functional diversity. *BioScience* 41:470–474.

Steele, J. H., S. Carpenter, J. Cohen, P. Dayton, and R. Ricklefs. 1989. *Comparison of Terrestrial and Marine Ecological Systems: Report of a Workshop Held in Santa Fe, New Mexico.* Report prepared by the Steering Committee.

Stegeman, J. J., P. J. Kloepper-Sams, and J. W. Farington. 1986. Monooxygenase induction and chlorobiphenyls in the deepsea fish *Coryphaenoides armatus*. *Science* 231:1287–1289.

Stehli, F. G., R. G. Douglas, and N. D. Newell. 1969. Generation and maintenance of gradients in taxonomic diversity. *Science* 164:947–949.

Stehli, F. G., and J. W. Wells. 1971. Diversity and patterns in hermatypic corals. *Systematic Zoology* 20:115–126.

Stoecker, D. 1992. "Animals" that photosynthesize and "plants" that eat. *Oceanus* 35:24–27.

Stone, R. 1992. Swimming against the PCB tide. *Science* 255:798–799.

Stoner, A. 1983. Pelagic *Sargassum:* Evidence for a major decrease in biomass. *Deep-Sea Research* 30:469–474.

Sugimoto, T., and K. Tadokoro. 1997. Interannual-interdecadal variations in zooplankton biomass, chlorophyll concentration, and physical environment in the subartic Pacific and Bering Sea. *Fisheries Oceanography* 6:74–93.

Taylor, P. 1993. A state of the marine environment: A critique of the work and role of the Joint Group of Experts on Scientific Aspects of Marine Pollution. *Marine Pollution Bulletin* 26:120–127.

Tegner, M., L. Basch, and P. Dayton. 1996. Near extinction of an exploited marine invertebrate. *TREE* 11:278–279.

Thorne-Miller, B. 1992. The LDC, the precautionary approach, and the assessment of wastes for sea-disposal. *Marine Pollution Bulletin* 24:335–339.

Tiner, R. 1984. *Wetlands of the United States: Current Status and Recent Trends.* U.S. Department of the Interior, Washington, D.C.

Toggweiler, J. R. 1988. Deep-sea carbon: A burning issue. *Nature* 334:468.

Toggweiler, J. R. 1989. Are rising and falling particles microbial elevators? *Nature* 337:691–692.

Toggweiler, J. 1994. The ocean's overturning circulation. *Physics Today* 11:45–50.

Travis, J. 1993. Probing the unsolved mysteries of the deep. *Science* 259:1123–1124.

Turner, S. M., et al. 1988. The seasonal variation of dimethyl sulfide and dimethylsulfoniopropionate concentrations in nearshore waters. *Limnology and Oceanography* 33:364–375.

Underwood, A. J., and E. J. Denley. 1984. Paradigms, explanations, and generalizations in models for the structure of intertidal communities on rocky shores. Chap. 11 in D. R. Strong, Jr. (ed.), *Ecological Communities: Conceptual Issues and the Evidence.* Princeton University Press, Princeton, N.J., pp. 151–180.

UNEP (United Nations Environment Programme). 1995. *Global Programme of Action for the Protection of the Marine Environment from Land-Based Activities.* Note by the secretariat, Intergovernmental Conference to Adopt a Global Programme of Action for the Protection of the Marine Environment from Land-Based Activities, Washington, D.C., 23 October–3 November 1995. No. UNEP(OCA)/LBA/IG.2/7, 5 December. UNEP, Nairobi, Kenya.

U.S. Congress. Committee on Merchant Marine and Fisheries. 1988. *Coastal waters in jeopardy: Reversing the decline and protecting America's coastal resources.* Oversight report of the Subcommittee on Fisheries and Wildlife Conservation and the Environment and the Subcommittee on Oceanography of the

Committee on Merchant Marine and Fisheries. U.S. Government Printing Office, Washington, D.C.

U.S. Congress. Office of Technology Assessment. 1986. *Serious Reduction of Hazardous Waste.* OTA-ITE-317. U.S. Government Printing Office, Washington, D.C.

U.S. Congress. Office of Technology Assessment. 1987a. *Technologies to Maintain Biological Diversity.* OTA-F-330. U.S. Government Printing Office, Washington, D.C.

U.S. Congress. Office of Technology Assessment. 1987b. *Wastes in Marine Environments.* OTA-O-334. U.S. Government Printing Office, Washington, D.C.

Van Dover, C. L. 1996. *Deep-Ocean Journeys: Discovering New Life at the Bottom of the Sea.* Addison-Wesley, Helix Books, Reading, Mass.

Van Dyke, J. M. 1993. Protected marine areas and low-lying atolls. Chap. 16 (pp. 214–228) in J. M. Van Dyke, D. Zaelke, and G. Hewison (eds.), *Freedom for the Seas in the Twenty-First Century: Ocean Governance and Environmental Harmony.* Island Press, Washington, D.C.

Venrick, E. L. 1982. Phytoplankton in an oligotrophic ocean: Observations and questions. *Ecological Monographs* 52:129–154.

Venrick, E. L. 1990. Phytoplankton in an oligotrophic ocean: Species structure and interannual variability. *Ecology* 71:1547–1563.

Venrick, E. L., J. A. McGowan, D. R. Cayan, and T. L. Hayward. 1987. Climate and chlorophyll *a*: Long-term trends in the central North Pacific Ocean. *Science* 238:70–72.

Vermeij, G. J. 1978. *Biogeography and Adaptation: Patterns of Marine Life.* Harvard University Press, Cambridge, Mass.

Vermeij, G. J. 1991. When biotas meet: Understanding biotic interchange. *Science* 253:1099–1104.

Villareal, T., M. Altabet, and K. Culver-Rymsza. 1993. Nitrogen transport by vertically migrating diatom mats in the North Pacific Ocean. *Nature* 363: 709–711.

Vitousek, P. M., H. A. Mooney, J. Lubchenco, and J. M. Melillo. 1997. Human domination of Earth's ecosystems. *Science* 277:494–499.

Wallace, R.L. (comp.). 1997. *The Marine Mammal Commission Compendium of Selected Treaties, International Agreements, and Other Relevant Documents on Marine Resources, Wildlife, and Environment.* Vols. 1, 2, and 3 and First Update. Marine Mammal Commission, Bethesda, Md.

Warwick, R., and K. Clarke. 1995. New "biodiversity" measures reveal a decrease in taxonomic distinctness with increasing stress. *Marine Ecology Progress Series* 129:301–305.

Warwick, R. M., and Ruswahyuni. 1987. Comparative study of the structure of some tropical and temperate marine soft-bottom macrobenthic communities. *Marine Biology* 95:641–649.

Waters, T. 1995. The other grand canyon. *Earth* 12:44–51.

Watts, M., R. Etter, and M. Rex. 1992. Effects of spatial and temporal scale on the relationship of surface pigment biomass to community structure in the deep-sea benthos. Pp. 245–254 in G. T. Rowe and V. Pariente (eds.), *Deep-Sea Food Chains and the Global Carbon Cycle.* Kluwer Academic Publishers, Dordrecht, Netherlands.

Weber, P. 1993. *Abandoned Seas: Reversing the Decline of the Oceans.* Worldwatch Institute, Washington, D.C.

Wecker, M., and D. M. Wesson. 1993. Seaborne movements of hazardous materials. Chap. 15 (pp. 198–213) in J. M. Van Dyke, D. Zaelke, and G. Hewison (eds.), *Freedom for the Seas in the Twenty-First Century: Ocean Governance and Environmental Harmony.* Island Press, Washington, D.C.

Weiss, R. 1988. Plentiful plankton noticed at last. *Science News* 134:68.

Weiss, R. F., J. L. Bullister, R. H. Gammon, and M. J. Warner. 1985. Atmospheric chlorofluoromethanes in the deep equatorial Atlantic. *Nature* 314: 608–610.

Wells, J. W. 1957. Coral reefs. Chap. 20 (pp. 609–631) in J. W. Hedgpeth (ed.), *Treatise on Marine Ecology and Paleoecology.* Memoir 67, vol. 1. Geological Society of America, New York.

Wells, S. M., and J. G. Barzdo. 1991. International trade in marine species—Is CITES a useful control mechanism? *Journal of Coastal Management* 19: 135–142.

Wiegert, R. G., and L. R. Pomeroy. 1981. The salt-marsh ecosystem: A synthesis. Chap. 10 (pp. 218–239) in R. Pomeroy and R. G. Wiegert (eds.), *The Ecology of a Salt Marsh.* Springer-Verlag, New York.

Wilder, R. 1997. Learning to cooperate: Joining science and policy for ecosystems governance. Paper presented at conference "California and the World Ocean '97." March 24–27, 1997, San Diego, Calif.

Wilkinson, C., and B. Salvat. 1997. The Global Coral Reef Monitoring Network—Communities, governments, and scientists working together for sustainable management of coral reefs. Paper prepared for the "Conference on Coral Reefs, An Associated Event of the Fifth Annual World Bank Conference on Environmentally and Socially Sustainable Development." October 9–11, 1997, Washington, D.C.

Williams, N. 1998. Overfishing disrupts entire ecosystems. *Science* 279:809.

Williamson, M. 1997. Marine biodiversity in its global context. Chap. 1 (pp. 1–17) in R. Ormond, J. Gage, and M. Angel (eds.), *Marine Biodiversity: Patterns and Processes.* Cambridge University Press, Cambridge, England.

Wilson, E. O. 1984. *Biophilia.* Harvard University Press, Cambridge, Mass.

Wilson, E. O. (ed.). 1988. *Biodiversity.* National Academy Press, Washington, D.C.

Wilson, E. O. 1993. *The Diversity of Life.* W.W. Norton and Company, New York.

Wilson, P., and D. P. Wheeler. 1997. *California's Ocean Resources: An Agenda for the Future.* California State Capitol, Sacramento.

Wilson, R., and E. A. C. Crouch. 1987. Risk assessment and comparisons: An introduction. *Science* 236:267–270.

Windom, H. L. 1991. *GESAMP: Two Decades of Accomplishments.* International Maritime Organization, London.

Winston, J. E. 1990. Life in Antarctic depths. *Natural History* 9:70–75.

Wirth, T. E. 1995. Values and political will. Pp. 29–31 in I. Serageldin and R. Barrett (eds.), *Ethics and Spiritual Values: Promoting Environmentally Sustainable Development.* Environmentally Sustainable Development Proceedings Series, No. 12. World Bank, Washington, D.C.

Woodwell, G. M. 1990. The earth under stress: A transition to climatic instability raises questions about patterns of impoverishment. Chap. 1 in G. W. Woodwell (ed.), *The Earth in Transition.* Cambridge University Press, Cambridge, England.

Working Group on the Protection of the Arctic Marine Environment. 1996. *Report to the Third Ministerial Conference on the Protection of the Arctic Environment, 20–21 March, 1996, Inuvik, Canada.* Ministry of Environment, Oslo, Norway.

World Commission on Environment and Development. 1987. *Our Common Future.* Oxford University Press, Oxford, England.

World Resources Institute. 1987. *World Resources 1987.* Basic Books, New York.

World Resources Institute, International Institute for Environment and Development, and United Nations Environment Programme. 1988. *World Resources 1988–89*. Basic Books, New York.
Wynne, B., and S. Mayer. 1993. How science fails the environment. *New Scientist*. June 5, 33–35.
Yentsch, C. 1990. Estimates of "new production" in the mid-North Atlantic. *Journal of Plankton Research* 12:717–734.
Zaitsev, Y. P. 1992. Recent changes in the trophic structure of the Black Sea. *Fisheries Oceanography* 1:180–189.

Index

About the Author

Boyce Thorne-Miller is senior scientist with SeaWeb, a marine environment communications organization, where her principal role is translating scientific information into language meaningful to the general public and applying science to ocean policy issues. She is a marine biologist who has worked in the environmental field since 1988, starting with the Oceanic Society, followed by Friends of the Earth, and Ocean Advocates. She currently serves on the boards of directors of the Arctic Network, Coast Alliance, and Ocean Advocates.

Boyce studied marine biology at the University of Washington and the University of North Carolina and oceanography at the University of Rhode Island, where she received a master of science degree. As a research associate at the University of Rhode Island, she did research on phytoplankton, seaweeds, and submerged aquatic vegetation. She also was on the teaching faculty of Rollins College in Florida and the University of Rhode Island Continuing Education Program.

During the past decade, Boyce has been active in promoting the conservation of marine biodiversity and environmental issues having to do with marine pollution and marine aquaculture.

Boyce is the author or coauthor of two previous books on marine biodiversity: the first edition of *The Living Ocean,* published by Island Press, and *Ocean,* published by Collins Publishers.